FEEDBACK

Uncovering the Hidden Connections between Life and the Universe

NICHOLAS R. GOLLEDGE

Prometheus Books
Essex, Connecticut

 **Prometheus Books**

An imprint of Globe Pequot, the trade division of
The Rowman & Littlefield Publishing Group, Inc.
4501 Forbes Blvd., Ste. 200
Lanham, MD 20706
www.rowman.com

Distributed by NATIONAL BOOK NETWORK

British Library Cataloguing in Publication Information Available

Library of Congress Cataloging-in-Publication Data Available

ISBN 9781633889330 (cloth : alk. paper) | ISBN 9781633889347 (epub)

♾️™ The paper used in this publication meets the minimum requirements of
American National Standard for Information Sciences—Permanence of Paper for
Printed Library Materials, ANSI/NISO Z39.48-1992.

For Frank and Evelyn.
May the feedbacks in your lives always be positive.

Feedback (*noun*):
the modification or control of a process or system by its results or effects*

---

* www.google.com/search?q=feedback+definition

CONTENTS

EARTH   1

Great God! This is an awful place.

—CAPTAIN ROBERT FALCON SCOTT, ON ARRIVING
AT THE SOUTH POLE, JANUARY 17, 1912

Earth had a difficult start. Around four-and-a-half billion years ago, barely a few tens of millions of years into its early life, our planet collided with another, half its diameter. Enough energy was released from the impact that part of Earth as well as that of the smaller body, Theia, was vaporized. Debris from both planets, ejected into space, coalesced to give us our moon.

For hundreds of millions of years our home was continuously assaulted by extraterrestrial impacts, yet from the tempestuous fury of the unrelentingly bombarded early Earth emerged everything that has been and that will ever be. The gargantuan feat of self-organization that brought us to where we are now spans a time scale so unfathomably vast that the only surviving remnants of Earth's fierce origins are ancient crystals entombed in rocks laid down a billion years later.

How did this infernal early world become the Earth we know today? Where did *we* come from? Evidence suggests that the same processes giving rise to plate tectonics on that early Earth also controlled the later emergence of life and the rhythm of the evolving climate. And that remarkably, right now, those very same processes also dictate the beating of our hearts and the vibrance of our societies. Why?

All of these things, and many more besides, are connected by common ways of working, unseen changes that bring about gradual improvements by refining, little by little, the way a system works. Adapting and evolving through an incalculable number of tiny adjustments, the enormous diversity of the processes taking place all around us, all of the time, have something in common. And those common behaviors are their *feedbacks*.

The feedbacks that underpin every aspect of our daily lives often lie hidden from view, running quietly in the background while we "get on with things." But if we look carefully, we can see them and identify within them a beautiful simplicity and elegance. Embarking on a journey to explore these feedbacks, a new world opens up to us, a world in which the fundamental essence of our existence and the intimate connections that run through our lives is revealed.

To understand feedbacks is to understand ourselves, our place in the world, our relationships with one another, and the ways we conceive of— and shape—our futures. In the pages that follow, we paint a picture, one layer at a time, a portrait in which we are the subject, nature is our muse, and our planet the canvas. We draw together fragmentary filaments of seemingly disconnected data and weave them together in a tapestry whose shadows of uncertainty demand from us as much attention as our subjects more brilliantly illuminated. Let us now, then, first sketch the envelope for our work and begin at the beginning, when our Earth was young.

Hell on Earth

The almost identical chemical composition of Earth and Theia reflects the depth of impact achieved during that initial collision.* Later in this chapter we discover how the formation of the moon led to feedbacks that not only controlled the physical development of the Earth, but also underpinned the processes that gave rise to the first atmosphere and to the origins of biotic life. But from those first moments, the most enduring consequence of this giant impact was large-scale melting of Earth's rocky mass and the formation of oceans of molten rock.

This eon of geological time, the Hadean, describes the first five hundred million years of our Earth's existence and takes its name from the Greek god of the underworld and of that hidden realm itself. Hades was a cold and dark place, a shadowy and gloomy cavern from which few mortals could escape. The thirteenth-century religious philosopher and

---

* One recent study suggests that the chemical similarities indicate a two-stage impact of lower velocity. Emsenhuber, A., Asphaug, E., Cambioni, S., Gabriel, T. S. & Schwartz, S. R. Collision chains among the terrestrial planets. II. An asymmetry between Earth and Venus. *The Planetary Science Journal* 2, 199 (2021).

poet Dante Alighieri subsequently built his allegorical *Comedia* on similar foundations. Dante's descent into Hell takes the Florentine poet ever downward through a funnel-shaped pit of twenty-four great circles to the very center of the Earth. There he encounters the fallen seraph, Lucifer—the "Emperor of the sorrowful realm"—frozen to his waist in a lake of ice. Immobile and tormented by his loss of power, Lucifer is forever condemned to impotently beat his corrupted angelic wings. To Dante and to earlier Greek scholars, the idea that the center of the Earth, hidden and distant from the sun, should be cold and devoid of God's light was intuitive and logical. And in the beginning, the aggregating minerals that condensed from the swirling cloud of gas and dust that formed our primordial solar system brought forth a planetesimal that was perhaps relatively cool and inert compared to its present state. But the continual bombardment of our planet by meteorites quickly swelled the early Earth, increasing its gravitational field and pulling in more and more debris that solidified into an increasingly dense planet. Heat from those meteorite impacts warmed the planet, and as it grew larger, the sinking of heavier metallic elements released energy that raised the Earth's temperature by a thousand degrees. A hot, dense iron and nickel core began to form, triggering thermal convection in the layer of molten rock above. The decay of radioactive elements increased and allowed for a constant supply of heat inside the new planet. Today the inner core exists as a solid[*] sphere of iron and nickel, five thousand kilometers beneath the surface of the Earth and heated by intense pressure to seven thousand degrees, hotter than the surface of the sun.[†]

During the first two hundred million years of this early era, the impact frequency of the asteroids bombarding Earth declined, but geological studies based on the scant surviving evidence, as well as computer modeling, indicate that this period of bombardment was instrumental in determining how the outer crust evolved. The crust is part of the lithosphere, the solid outer layer of the Earth through which heat flows mainly by conduction, a slow and inefficient process. The lithosphere therefore provides a strong insulating layer, a lid over the hotter asthenosphere and softer part of the mantle below. Since the middle of the twentieth century, we have accepted the idea that lithospheric plates on the surface of the Earth migrate over time, but as far as we know, our planet is the only one in our solar system to

---

[*] A recent study, based on the way that seismic waves travel through the Earth, suggests that the inner core may in fact have patches that aren't completely solid. Butler, R. & Tsuboi, S. Antipodal seismic reflections upon shear wave velocity structures within Earth's inner core. *Physics of the Earth and Planetary Interiors* 321, 106802 (2021).

[†] The sun's surface is about 5,000 kelvin. Its gaseous and much less dense corona, however, is around 2,000,000 kelvin.

have such an active surface. The default, it seems, is for "stagnant" lids—an outer casing that is neither fractured nor mobile. Why is Earth so special? What was it that broke our shell, and what feedbacks did this produce? Some theorists propose that it was the period of intense asteroid bombardment, early in the Hadean, that triggered tectonic activity on the early Earth.[1] As with the collision of Theia, other large impacts would have created thinner and hotter patches within the surface rocks that encouraged the hot and viscous rock below to slowly rise to the surface. Simulations show that these large-scale upwellings would have been initiated directly beneath impact sites and that as these newly formed "plumes" of ascending magma rose upward they entrained other smaller plumes, organizing the internal structure of the Earth in a way that localized the upwellings to specific regions. Above the zones of upwelling, these plumes spread outward in horizontal flows that dragged at the solid crust above them.

This inner turmoil marked the beginning of the long evolution of the Earth's interior, but what of its atmosphere, the more visible realm above the Hadean underworld? This was a time of relative darkness with the young sun only 70 percent as bright as today, but it was also a time of intense heat. A heavy atmosphere rich in carbon dioxide produced a surface pressure twenty-seven times greater than that of today, allowing liquid water to form at the surface despite temperatures of more than 200 degrees Celsius. This carbon dioxide came from the slowly cooling magma oceans early in the Hadean; once released into the atmosphere, it produced an intense greenhouse warming of the primitive climate. More carbon dioxide in the atmosphere led to more intense and frequent lightning flashes than today, and remarkably, these fleeting events may have played a pivotal role in producing the elements needed to fuel early life. Lightning striking the ground heated clay minerals to 3,000 kelvin, a thousand degrees higher than their melting point, forming fulgurites—silica glasses that also incorporated traces of the elements nickel and phosphorus. Weathering of these fossilized storm rocks provided a continual supply of the building blocks needed for molecules such as DNA. And with the rain falling from the Hadean skies incorporating more of the abundant carbon dioxide than it does today, the resulting carbonic acid accelerated the processes that chemically alter and erode rock, allowing fertile sediments to accumulate. At first this acid rain also lowered the pH of the primordial oceans, but gradually the carbon that was first released into the atmosphere from the mantle, and then returned to the ocean by rain, was progressively sequestered once more in the mantle, lowering the atmospheric concentration of carbon dioxide and allowing this superheated world to cool.

Did this complex chain of feedbacks, this fortuitous consequence of an extraterrestrial collision that triggered mantle heating and convection, set the stage for the emergence of life on Earth? Or were there other hidden connections that helped organize, adapt, and optimize this cycle? To answer this and to understand the significance and consequence of these first feedbacks, we need to inquire more deeply into those mantle plumes, plate tectonics, and the birth of continents.

Controlling the Chaos

Cracks in molten rock heal quickly, so before the outer shell of the Earth could form rigid, breakable plates, it needed to cool. But solidification itself still wasn't enough to allow a patchwork of capricious rock slabs to develop, migrate, and regenerate themselves in a self-sustaining and ever-cycling way. Five hundred million years into the life of the early Earth, the rocks deep below the surface were still two hundred degrees hotter than they are today. The rocky surface was also too hot, and that made it buoyant, unable to sink into the more fluid rock beneath. Plate tectonics, it seems, is a rather fussy phenomenon. That outer shell needed to cool just enough to become denser than the hotter mantle below and so allow it to sink. Upwelling mantle plumes pushed and pulled it sideways, and together these forces enabled the fractured outer skin to sink downward at its now cool-enough-to-crack margins. And that cooling was a slow process. In the course of the fifteen hundred million years of the Archean eon, from four to two-and-a-half billion years ago, the Earth went through a gradual evolution from a hot phase, in which the "stagnant lid" was too buoyant, to a cooler phase in which the now "squishy lid" was just about dense enough to allow episodic and localized sinking and the beginnings of the tectonic processes that later became widespread.[2] But how, and why, did Earth's almost unique plate tectonics get started?

Although there is still no consensus, there are, of course, several theories. One leading argument suggests that the force of upwelling relatively hot rock in the mantle eventually became strong enough to break the increasingly cool and brittle surface layers. That would allow these newly formed "plates" to move, but even if rising plumes of hot rock had thinned and weakened the crust, these plates would quickly collide with one another. What was to stop them recombining? The secret to successful plate tectonics lies in these junctions, the boundaries where plates meet or separate. For effective plate movement, a new process was needed—subduction. Where one colossal slab of Earth's surface encountered another, subtle differences in their buoyancy could mean the difference between life and death; that is, a colder, denser, less buoyant slab would tend to sink beneath a lighter one. Pushed by gravitational forces acting on the dense plates, and helped along from afar by the outward-spreading hot rocks rising in the mantle plumes, the cooler margins of Earth's plates sank slowly back into the interior from whence they had come. But these gigantic monoliths grating against one another would rapidly seize up as the friction of their communion increased. To maintain the plate tectonic process, therefore, a subduction zone needs to be lubricated, either by water within the rock or by soft sediments overlying the rock that are taken down with the subducting plate. And it isn't just the surface whose friction could put an end to this slow-motion cascade. Beneath the dense lithosphere, a layer of water or more fluid melt helped ease the passage of its flow over the warmer aesthenosphere below.[3] Overcoming friction in this way, the sinking rock slabs exerted a suction force that pulled still-buoyant parts of the plate, while the subducted rock itself eventually melted, to be recycled through the mantle to feed a distant rising plume. Dating the beginnings of plate tectonics is therefore largely a question of dating the onset of subduction. And although the task of dating melting rocks might seem impossible, geologists now think that the earliest evidence of subduction dates from the early Archean, between three-and-a-half and four billion years ago. But at that time, it was just a local phenomenon; for it to become the globally connected process we recognize today, more time and more feedbacks, were needed.

Today, the absence of oceanic crustal rocks older than about two hundred million years tells us something of the average recycling timescale for tectonic plates.[4] And as it turns out, the very process of recycling can help pinpoint when large-scale plate tectonics became pervasive. This is because the early Earth would have been rich in specific isotopes of the chemical elements that were abundant during the first tens of millions of years after the solar system came into being. In a mantle without recycling, those isotopes would be preserved, locked into rocks at the surface of the Earth. Plate tectonics, however, mixes those old isotopes with others already present in the mantle, diluting their abundance. Rocks forming at the surface during the transition from a stagnant to mobile lid would, therefore, show a drop in these marker isotopes coincident with the onset of mantle mixing and plate movement above.[5] And these markers point to a time around three billion years ago as the period when this transition took place, setting in motion a perpetual process of creation and destruction that even today continues to adapt and shape our world.

The petrified shadows we now study tell stories of the hidden reality of the underworld, whispering secrets of their past just loud enough that we might descend like Dante and bear witness to the machinations of the inferno beneath our feet. The vehicle of our descent, however, is no longer religious vision. Instead we are guided by the murmurs of the Earth, seismic events whose waves travel through the deepest circles of our planet and imprint telltale traces of their journey on geophysical equipment arrayed across Earth's surface. But what clues do they provide to us, the assembled eavesdroppers on their silent communications? For mantle plumes to play any role in initiating plate tectonics, these rising conduits of magma would need to be monstrously proportioned, growing ever larger as they scavenge molten rock from one another. But their heat means that seismic waves avoid them, instead finding faster routes through colder rocks at their margins. So the eavesdroppers map the waves to reveal the negative space between, the shadows that expose themselves by trying to hide. And these shadows, these slow zones, show something tremendous, for they map out hidden structures so gigantic that they stretch from the very lowest layer of the mantle to the very surface of the Earth almost three thousand kilometers above. Titanic in stature, these magmatic "trees"[6] feed on heat from Earth's core, funneling magma vertically through a trunk a thousand kilometers across, toward branches that narrow as they reach for the lithosphere above, and into the terrestrial volcanoes they nourish. Geophysical exploration has unmasked a subterranean realm of heat and fury whose gateways are volcanic edifices like Réunion, one of the most active volcanoes in the world. "Down we must go, to that dark world," wrote Dante, as he "stood on the steep brink whereunder runs down the dolorus chasm of the Pit, ringing with infinite groans like gathered thunder." Dante had envisioned a "huge funnel-shaped pit,"[7] narrowing to the center of the Earth; now, seven hundred years later, scientists might just have proved him right.

In the gravitational sinking of cooling and increasingly dense lithospheric plates and in the mantle plumes beneath them, we have identified a driving force for our plate tectonic system, for our machine of perpetual motion. But how could these processes, even if sufficiently forceful to tear apart solid rock and fracture the uniform shell of the young Earth, coordinate their actions in such a way as to give rise to the intricate pattern of plates we see today? The Greeks believed that the primordial parents Uranus (sky) and Gaia (Earth) spawned Titans, gods possessed of incredible strength who ruled the Earth until Zeus and the Olympians banished them to the underworld of Tartarus, a dark abyss deeper and more hellish than the gloomy caverns of Hades. Are these gods, these mantle forces, responsible for tearing and separating slabs of Earth's shell, pitching them outward into zones of slow-motion collision where the loser is banished to Tartarus to be recycled and, hundreds of millions of years later, to be violently reborn?

Bits and bytes tell us something that perhaps the myths of the Greeks cannot. When computer simulations are used to hold a torch to this dark phenomenon, they reveal something magical emerging from the brutality of the Titanomachic struggles. Widespread cooling allowed Earth's outer layer to harden and become brittle, but as "heat pipes" within this shell slowly closed, the solid skin trapped more and more of the heat emanating from the warmer mantle below. Heating the shell caused it to swell and to crack. At first, the cracks took tentative steps from one site of weakness to another, sometimes meeting another fracture opening in the same direction. Emboldened in this way, the splitting shell developed longer ruptures, new ones forking from their parents at right angles or creating triple junctions, each time extending the damage and connecting different generations of breakages. At first gradually but then increasingly rapidly, the warmed and expanding single plate disintegrated through an avalanche of fracture events into randomly distributed but almost equally sized polygonal plates.[8] Once this fully connected global plate network was born, the rate of fracturing then dropped, and the process was complete. And the most beautiful thing about this process? Perhaps that this paroxysm of tectonic activity took place so quickly, in so orderly a fashion that it appeared somehow intentional? Or perhaps that it reveals an emergent process of self-organization taking place on a planetary scale? No, it is that the pattern produced from these titanic forces over millions of years is one in which there is no redundancy of effort, no wastage. The tessellating tiles of the gods are so perfectly designed, so optimally arranged, that the total length of the fractures produced is as short as it could possibly be.

The Birth of Nuna, the First Supercontinent

Episodic and localized shifting of Earth's lithosphere throughout the Archean gave rise to continents through a process of orogenic thickening. Typically, magma upwelling to the Earth's surface from the mantle produces "oceanic" rock composed of dense basalt, rich in magnesium and iron. At the margins of these oceanic plates, their sinking edges melt as they return to the mantle. Melting at relatively shallow depths allows the minerals in the rock to separate from one another due to differences in their density or saturation. Over time, this process of fractionation produces molten rock whose chemical composition is dominated by the elements silicon, aluminum, sodium, and potassium. These lighter elements make the molten rock relatively buoyant, allowing it to float above the descending basaltic melt, which is heavy with iron. At the same time, the compressive forces taking place at these margins thicken the new, light rocks, pushing them both upward to form mountains and downward to form "roots," or "keels."

The first continents rose above sea level in the early Archean, around two-and-a-half to three billion years ago, and because they were light and resistant to the subduction and recycling that condemned their oceanic counterparts to eternal punishment in the great circles of hell, some of these continental rocks still survive today. The longest-lived continental blocks are called cratons (from the Greek *kratos*, meaning "strong"), and these cratons comprise nearly two-thirds of the current continental landmass. Cratons are economically important today, for they contain all of the world's diamonds and more than 90 percent of its gold and platinum.[9] But during the Archean, they were even more important, for as these new lands reached heavenward, they were subject to the same wearing down by the elements as that which dissolved the fulgurites. Today we know from simulations using state-of-the-art computer models that rainfall in "hothouse" worlds like the Archean was quite different from that which occurred under cooler conditions. As the surface of the Earth warms above a critical threshold of around 45 to 50 degrees Celsius, something strange happens. Instead of radiating heat out to space, the lower troposphere (the lowest level of the atmosphere) starts to retain more heat, and because there is no longer as much of a difference in temperature between the Earth's surface and the air far above, the normal processes of convection break down. Instead, the air column becomes more strongly layered, and both heat and moisture are trapped in the layers closest to the surface. As the energy builds in the atmosphere, it slowly reaches a critical level where it can no longer stay put. All the pent-up frustration of the superheated, supersaturated air is suddenly released as it at last ascends to heights where it becomes cool enough to release its moisture[10] in episodic but torrential storms. In this hothouse regime, it never just rains; it pours.

Intense downpours generate far more surface runoff and are far more effective at eroding the land surface than more gradual rainfall, even if the total amount of water released is the same. As a result, these superstorms chemically altered and physically transported sediments more efficiently than is typical under today's climate, washing them back to the oceans in thundering rivers excitedly exploiting the new landscape. And these sediments appear to have played a critical role in the tectonic cycle by lubricating the downward subduction of sinking oceanic plates. At times when sedimentation into the deep oceanic trenches was greatest, subduction could proceed more easily.

Those rivers were effective at eroding and transporting sediments, but what if the falling rain were to freeze and settle to Earth as snow, causing great ice sheets to expand across the elevated continents? Soft sediment beneath sliding glaciers is eroded easily, particularly during times of ice advance and retreat, when subglacial sediment transport is most effective. Is it coincidence that the two largest subduction events followed global glaciations? One leading theory suggests not. According to Stephan Sobolev and Michael Brown,[11] these ice worlds not only accelerated plate tectonics, but in facilitating the movement of a globally connected network of plate boundaries, sediments allowed the gradual migration and eventual collision of nearly all of Earth's continental plates,[12] including those that now comprise parts of Siberia, northern Europe, parts of North and South America, and Australia. Simply put, the seemingly trivial feedback of allowing water to freeze over a sufficiently large area led to the formation, two billion years ago, of Nuna, the world's first true supercontinent.

The Life Cycle of the Earth System

In Nuna we are confronted with the tremendous amplifying power of the positive feedback cycle. At first slow and disparate, the movements of those first fragments of Earth's lithosphere cautiously pioneered a process, proving a pathway for others to follow. And with more certainty and vigor, those other fragments did just that, subducting and colliding, melting and separating, until a global network of plates built upward to such extent that the surface of the Earth was transformed. Our Earth exists at a distance from its nearest star that results in solar luminosity being neither too high nor too low for liquid water to dampen its skin. We live in the "habitable zone" of the solar system where the solar parameters are just right for life. But fortuitous orbital placement does not presuppose the establishment of habitable conditions; for that, we need a favorable atmosphere. Earlier we saw how the Hadean atmosphere, dense with carbon dioxide, allowed water to form at the surface despite temperatures twice that which would rapidly boil away the oceans today. And this water, raining from the suffocating sky, transformed silicate rocks to carbonates that locked away carbon dioxide in oceanic trenches to be unhurriedly returned to the mantle. This cycling of carbon through the atmosphere, ocean, and solid Earth reveals the intimacy of connections between the realms of Zeus, Poseidon, and Hades, and it is this coupling that allowed our planet to make the most of its position in the zone of habitability. By sequestering carbon dioxide from the atmosphere, the solid Earth imposed control on its climate like a thermostat. Plate tectonics produced fresh and weatherable rock at the surface, which drew down carbon dioxide, cooling the climate. But cool the climate too much and liquid water would freeze, inhibiting life. Glacially enhanced subduction therefore provided the perfect escape mechanism for this standoff, promoting faster recycling and the return of carbon dioxide to the atmosphere through volcanic activity.

Where subducting slabs sink into the mantle, they carry with them water and other volatiles. These in turn control the production of magma, the formation of continental crust, locations of mineral abundance, and also the occurrence of earthquakes, because escaping water lowers the stresses on geological faults.[13] Subducting margins are typically associated with volcanic arcs, chains of volcanoes hundreds of miles long bursting through Earth's surface where rising magma can most easily find a path. Distinct from volcanoes overlying plumes, such as Réunion, volcanic arcs overlie oceanic trenches. These are the most dynamic zones on the surface of the Earth, and it is no surprise that where plates are pushed downward, scraping roughly against one another as they battle for buoyancy, they not only release melt to feed volcanic arcs, but also trigger swarms of earthquakes in precisely the same regions.

It makes sense therefore to see such events occurring together in regions where subduction takes place, but until recently it was believed that large earthquakes and volcanic eruptions were unconnected at the global scale. In the last few years, however, a brave new theory has been put forward that challenges this very basic tenet of geophysical assumption. Analyzing the records of large earthquakes and explosive volcanic activity over the last sixty years, a team of researchers in the United Kingdom found something quite unexpected.[14] Rather than these two phenomena being independent of one another, they found the opposite—periods of high global seismic activity tended to coincide with periods of high global volcanic activity. Statistical tests showed that the odds of this correlation occurring by chance were extremely low. There are well-understood mechanisms to explain how the abrupt changes in stress that take place during an earthquake can either trigger nearby seismic or volcanic activity, but it is not clear how these connections could exist over thousands of kilometers, because such stresses decay rapidly with distance. Maybe, the authors suggested, these changes, if repeated over and over, might be sufficient to incrementally disturb magmatic systems deep inside the Earth. Just like the slow buildup and rapid release in the rainfall patterns on early Earth, these results hint at a feedback operating at a global scale that couples two distinct phenomena, synchronizing them in a planetary waltz whose rhythm we're only just beginning to feel.

Through their coupling of tectonics, weathering, and volcanism, our brotherly Greek triumvirate, Zeus, Poseidon, and Hades, kept one another in check so that neither extreme greenhouse nor icehouse climates found permanence. But there is still one other actor whose face we've not yet seen—magnetism. Magnetic rocks, or lodestones, were known to the Greeks from at least the sixth century BC when Thales of Miletus, according to the later writings of Aristotle, asserted that these stones possessed a soul, for they exerted some motive power over nearby iron objects. A century or so later, Empedocles conceived that magnets had pores, like living things, and that these pores were somehow receptive to the "effluences" from iron. But what they can't have known was that it is because of magnetism that the early Earth was able to retain its primordial atmosphere. The solar wind, a torrent of high-energy particles ejected from flares at the sun's surface, can easily strip the atmosphere of a planet by sweeping away ionized gases. But a strong planetary magnetic field repels this wind, pushing it further from the upper parts of the atmosphere and so limiting the loss of atmospheric gas.

Without Earth's magnetic field, not only would our atmosphere be quickly eroded, but the exposed planetary surface would then be subject to harmful radiation. Under such conditions, life would struggle to gain a foothold. As far as we know, Earth has maintained an internally generated magnetic field throughout much, if not all, of its history. This is unusual, for it seems that of all the other terrestrial bodies in our solar system only Earth, Mercury, and Ganymede, one of Jupiter's seventy-nine moons, possess a magnetic field. Of these three, only Earth has a substantial atmosphere, which in turns means it is the only one of the three able to retain heat during periods of darkness and thereby stabilize daily temperature changes and allow liquid water to exist at the surface. Magnetism makes our world livable, but how does the Earth retain its magnetism? For that, we have to thank plate tectonics. For the continued overturning and cycling of the mantle brings cooler surface rocks downward and moves molten rock, heated by the core, upward to the surface. This cycle maintains a high flux of heat outward from the core, which in turns favors convection-driven overturning in the iron-rich molten outer core. And convection in this large, rotating body of electrically conductive liquid acts as a dynamo. The more effective the plate movements above, the more efficient the dynamo within. In short, it was plate tectonics that first allowed our atmosphere to evolve, and it was plate tectonics that then powered the magnetism that prevented its loss.

If Thales's conjecture held true—that magnetic rocks embodied something beyond the physical—then our rocky planet itself must equally qualify for such an attribute. The Greek philosopher Plato (428–348 BCE) conceived of a human soul[15] that transcended space and time and persisted without end, forever casting off one incarnation only to adopt another. Might the Earth itself also conspire in such cycles of death and rebirth? Geological evidence indicates that our magnetic soul has come and gone nearly two hundred times during the last eighty million or so years, each time weakening to only a few percent of its present-day strength and then regrowing in vigor but with its alignment, or polarity, reversed. Magnetic reversals and variations in the strength of the field reflect deep Earth processes taking place over hundreds of thousands to hundreds of millions of years, yet despite concerted analysis, it remains a mystery as to why or how these episodes are triggered.

Though the vascillations of our planetary soul are then apparently chaotic and unpredictable, our physical body undergoes more regular cycles of decay and rejuvenation. Ice ages during the Archean accelerated the erosion of sediment from continental areas and lubricated the subduction of oceanic plates, giving birth to Nuna, our first true supercontinent. Since then, the fragments of our shattered shell have cavorted and whirled, swaggered and sashayed, strutted and flounced their way in imperceptibly slow motion to the counterpoint of mantle plume cadence and subduction zone slippage. From this stop-motion animation of tectonic reorganization emerges the beat of a hidden heart. Cyclic coalescence brought our cratons together into supercontinent formations every four to five hundred million years, each time juxtaposing old friends with new neighbors, adding, subtracting, and modifying pieces of a vast geologic jigsaw that most recently gave us Pangaea, the mother of our modern lands. Two hundred million years from now a new supercontinent, Aurica, will form as the Atlantic and Pacific Oceans close and a new ocean splits Asia in two. These ocean basins thrum to the meter of their own cycles, shorter but superimposed upon that of the continents. Together these *danseuses étoiles* give rise to a supertidal cycle, rising and falling in strength as the interference of the continental outlines on the lunar pull of the oceans waxes and wanes.[16]

These formidable restructurings of the surface of the Earth lay a foundation over which operate processes whose pulse beats more quickly. Records are sparse as we go further back in time, so gathering sufficient data to confirm how quickly or frequently these changes occur is challenging. Not only that, but as we look more and more closely at shorter and shorter timescales, we might see more detail but the scale of its significance might be less clear. A detailed archive chronicling the demise of a certain species of marine plankton might be valuable for reconstructing ecological changes in the past, but what can it tell us of global mass extinctions? The landscape of discovery for such information is a mosaic of both fortuitous uncoverings as well as directed investigations, fragmentary in time, space, and disciplinary realm. Hidden connections often remain well hidden. But as with all jigsaws, we know that the pieces must fit together somehow, even if the image is far from finely depicted. To scrape value from this dark cave of imperfectly revealed shadows requires both the robust application of the best scientific methods and enough willingness to see what *might* be there, lurking furtively in the negative space, the spaces between the known. Because something deeper exists there, in the openness to connect the incomplete. It is the humility to accept imperfect knowledge and the optimism to ask the question, *what if . . . ?*

Over the last four decades, data from nearly a hundred geological events spanning the last 260 million years have been published and analyzed in terms of the cyclicities they might reveal. Simply put, if you bring together all of the available evidence concerning phenomena such as the periodic comings and goings of species, the rise and fall of sea level over millions of years, or episodes of increased or decreased volcanic activity, do the data show a repeating pattern through time or are all these events completely random? Scientists are divided. Some researchers feel that the data are too sparse and too uncertain to determine any kind of regular cycles of change. But to others, ten peaks in activity can be teased out, perhaps suggesting a natural cycle length that averages around 27.5 million years.[17] Much of the strength of this apparent cyclicity is linked to extinctions of various species or groups of species. But more than half of these episodes might also coincide with pulses of volcanic activity that would have released sufficient carbon dioxide into the atmosphere to trigger abrupt and severe atmospheric warming and loss of oxygen from the global oceans.

Beyond our planetary boundaries, fluctuations in solar energy over the same kind of timescale may also be responsible for controlling events on Earth, changing our climate from hothouse to ice-age worlds and triggering feedbacks through continental erosion or sediment deposition in oceanic trenches. At an even larger scale our solar system itself wobbles in and out of the immense flat plane formed by the spiral arms of our galaxy, and this oscillation occurs with a thirty-million-year frequency. Ages of impact craters on Earth show that we were periodically bombarded by galactic debris in a way that varied over this thirty-million-year timeframe as our star navigated through denser regions of space. To some researchers, these data are clear, but others are not so sure because the mechanisms that could produce such cyclicities are not always obvious, or at least, not yet. Whatever the explanations, we are nonetheless faced with the intriguing and at the same time terrifying prospect that these cyclic episodes represent something, "periodic, coordinated, and intermittently catastrophic."[18]

Setting the Stage for the Emergence of Life

Because repeated environmental changes taking place over millions of years are poorly recorded in the geological record, they are hard to predict with any certainty. But those occurring over thousands rather than millions of years should, in theory, be easier to anticipate. In the Aegean Sea of the eastern Mediterranean region sits the island of Thera, or Santorini, an explosive volcano whose last eruption in around 1600 BCE was one of the largest volcanic eruptions on Earth in historic times. This late Bronze Age eruption destroyed the island itself and the Minoan settlement it hosted and triggered earthquakes and tsunamis that wiped out communities on neighboring islands and along the Cretan coastline. The volcanic plume and associated lightning are thought to have been witnessed as far away as Egypt, and the global cooling that resulted from the volcanic ash cloud thrown into the atmosphere coincides with a cold wave recorded by Chinese chroniclers. But this far-reaching volcanic explosion was not simply a one-off, random event that impacted unlucky bystanders. The eruption in 1600 BCE was the most recent event from the Santorini volcano in the second of two eruptive cycles stretching back two hundred thousand years. Over the most recent 224,000 years, 208 eruptions identified in geological sequences all correspond to times when sea level fell by forty to eighty meters below present levels.[19] As sea levels subsequently rose higher than minus-forty meters, volcanic eruptions stopped. Santorini now lies quiet, just beyond the ten-thousand-year window of activity following sea level rise above that critical threshold, and most likely on the cusp of beginning a third eruptive cycle. Unfortunately for the Minoans, however, the seas rose slightly too slowly to prevent catastrophe.

The feedbacks that controlled the rise and fall of the global ocean that regulated the eruptions of Santorini and the decimation of a Mediterranean civilization are these days relatively well understood: over hundreds of thousands of years, variations in the amount of sunlight reaching the Earth as our planet wobbles on its axis and as our orbit around the sun slowly stretches and contracts led to the growth of great ice sheets at high latitudes. The additional weight of ice pushed down on the land and added pressure to magma chambers beneath the crust. The additional pressure helped keep the magma in place. The ice sheets were growing in only some parts of the world, but by removing vast amounts of water from the ocean, these ice sheets allowed sea level to drop by more than a hundred meters all around the globe. In areas without ice sheets, the lowered sea levels reduced the pressure that previously pushed down on magma chambers far below. Slowly but surely the hot and buoyant rock migrated upward more easily than it could have done before. Rock that was once held hostage in solid form by the great weight above now melted as it rose and decompressed, fueling a resurgence in volcanic activity. And so it was that a seemingly minor change in the balance of sunlight at the surface of the Earth due to imperceptible orbital changes grew, through cascading positive feedbacks, into a global climate and sea level perturbation of such magnitude that it disrupted the inner workings of the Earth and triggered disastrous geophysical events that devastated, for one civilization at least, life at the surface.

Throughout Earth's history, our complex planetary system has undergone changes such as these in repeated and predictable ways, tracing out the same familiar paths each time. But despite the repetition, each new cycle differs from the last. Unpredictable or chance events taking place in different parts of the world, affecting different pieces of the system, lead to different outcomes each time a cycle of change takes place. And this allows the Earth system to explore alternative scenarios, to toy with the idea of "what if . . . ?" When we look back at these oscillations, the large-scale features of each new cycle may look similar to those that went before, but the details are altered. The coupled Earth system absorbs a flow of energy from the sun, disperses it through its component parts, and responds to the stimulus by altering its structures in ways that modify how it will absorb the same stimulus next time. Adapting through change, our Earth system optimizes itself as part of a cognitive cycle, a process of adaptation or learning. Without change as a key part of this cycle, we would stagnate in a stationary world, locked in a pattern from which no new behaviors, no novel opportunities, would arise. But by combining regular, cyclical variations with unpredictable, one-off events, our Earth system allows chance events to have meaning, to incrementally nudge our evolution one way or another, to explore a new path without throwing us completely off course.

The emergence of intelligent life and sophisticated cultures is proof that something meaningful did indeed take place in the four-and-a-half-billion-year-long cognitive process of Earth's evolution. And that something moved Earth's state further and further from its previous equilibrium and toward a trajectory in which myriad microscopic jitters disrupted and corrupted the previous way of things just enough that in each orbit of each cycle of change, our Gaia of the ancient Greeks has grown and matured, has become more and more clearly a portrait of the hidden beauty that binds and connects all things across space and time.

How then to construct this picture, to arrange our composition in the way that it most benefits from the direction of our illumination and to choose the pigments that will bring vibrance where warranted, solidity where needed, and clarity where detail is required? Through the words above we have readied our palette, outlined our envelope, and populated that shape with forms, lines, and shadows. Our canvas is the physical Earth, the ocean and atmosphere our oils. Now we must paint and see how our world arranged itself so that eventually life might emerge.

Making a House a Home

We are lucky. Perhaps uniquely so. Not only does our host planet exist within the habitable limits of our solar system, but the immeasurable and unpredictable sequence of feedbacks that has taken place since the formation of our primordial Earth has been conducive not just to the emergence of life, but to the proliferation and development of complex multicellular organisms and to sentient, introspective, and cultural beings. As far as we know, this process of biotic creation and evolution is unique in the universe. And yet the self-aware species of today's world are built from the same elemental building blocks—hydrogen, carbon, nitrogen, oxygen, phosphorus, and sulfur—that exist throughout space, so how did just the right combination of elements end up in just the right place, under just the right conditions?

Hydrogen and oxygen are both essential elements individually, but combined into water, they underpin all life. As a solvent, water is more able than almost any other liquid to dissolve substances into forms that allow chemical reactions to take place in living cells. Water conducts heat and is stable across the wide range of temperatures experienced at the surface of our planet. It allows light to penetrate and, in doing so, can support life within it. And this life, like all life on Earth and perhaps elsewhere if it exists, is carbon based. Carbon molecules are strong and stable, and their structure allows long chains to form. Because carbon atoms bond easily with hydrogen and oxygen, complex organic compounds such as carbohydrates, proteins, and fats can be assembled from relatively simple initial components. But for life we need complex proteins, built from amino acids; for organisms to replicate and evolve, they also need deoxyribonucleic acid (DNA), a molecule that can encode genetic information and pass it from one generation to another. Nitrogen is needed for both amino acids and DNA, but nitrogen in the atmosphere cannot be readily absorbed by plants, so first we need bacteria in soil or water to convert the nitrogen into a form that can be more easily assimilated into living cells. And these cells need a fifth element, phosphorus, to maintain their membranes and allow selective transfer of substances into and out of the cells. Phosphorus is also a key component of DNA and another organic substance, adenosine triphosphate (ATP), the power supply for nearly all energy-dependent cellular processes. Without phosphorus, organisms simply would not be able to live and grow. Finally, sulfur. In some extreme environments, such as deep ocean hydrothermal vents where light is absent and oxygen concentrations are low, bacteria flourish through a process of chemosynthesis, drawing energy from hydrogen and sulfur rather than using light to convert carbon dioxide into carbohydrates (*photo*synthesis) the way that plants do.

With these six elements, forged in stars at temperatures of hundreds of millions of degrees and jettisoned into the cosmos through supernovas, life could begin. But for continued growth and evolution, these elements need to be recycled, transformed, combined, and recombined into ever more complex compounds that adapt and change according to external driving forces and internal feedbacks. Deep in the early Archean oceans, four billion years ago, microorganisms such as bacteria were the first simple forms of life on Earth. Colonizing hydrothermal vents, feeding on elemental sulfur, and reducing it to hydrogen sulfide, these single-celled organisms would later become a source of food as more complex life became established.

Above this dark submarine world, the early Earth was bombarded by meteorites that delivered phosphorus along with other elements. But as these destructive bombardments waned in their frequency, it was lightning that created phosphorus directly at the Earth's surface, in the horizons of weathered rock slowly accumulating as plate tectonics thrust lighter, silica-rich, rocky continents further above sea level. Reactive phosphorus from fulgurites leached slowly into the first soils but in a way that did not disrupt the delicate processes that enabled these elements to support life.[20] The continual supply of storm-generated phosphorus, so critical for the DNA and ATP required by energy-dependent self-replicating organic life, was most abundant in tropical regions, where the warmer and wetter conditions allow life today to flourish most readily. As plate tectonics became globally established, driven by heat from the Earth's core and the titanic mantle plumes it fed, subducting plates dragged sediments downward into deep trenches. Heat from the mantle released water that was trapped in the rock, transforming its physical properties and liberating methane and other compounds that would nourish the increasingly populated waters above. The faster and more connected the processes of plate tectonics became, the more combinations, configurations, environmental niches, and opportunities were explored and exploited. The constant flow of energy from convective upwelling and gravitational sinking was dispersed through the plate tectonic system, and this system adapted and optimized itself into a pattern of organization that was the most efficient it could possibly be.

Evolution of life accelerated as the optimized solid Earth produced feedbacks that refined the atmosphere above. In forming continents—chemically distinct from the oceanic plates—tectonic processes reduced the flow of oxygen-absorbing minerals from the mantle to Earth's surface and allowed free oxygen from photosynthesizing blue-green algae to accumulate in the atmosphere. Planetary-scale changes to the dynamics of the Earth's interior thus led to a burst of atmospheric oxygenation 2.4 billion years ago, the Great Oxidation Event.

By continually colliding lithospheric plates, raising continents, and forming continental shelf areas with shallow seas, plate tectonics changed the ocean tides and the forces they exerted on the early Earth. Today, as it did then, our moon exerts a gravitational pull on our planet, stretching, deforming, then relaxing it as our daily rotation changes the angle of its pull. Our oceans lurch from one side of their basins to the other, and as they do so they drag imperceptibly on the seabed beneath, microscopically countering the rotational force of the Earth itself. By slowing our spin, our days get longer. A hundred years from now, a day on Earth will be 2.3 milliseconds longer than it is now.* But four billion years ago, when the incremental effects of this lunar forcing had yet to accumulate, the Earth's spin sped us through a typical day in only six hours.[21]

---

* http://hyperphysics.phy-astr.gsu.edu/hbase/Astro/tidfrict.html

By creating continents, plate tectonics allowed the slowing effect of this lunar tidal drag to be greatly enhanced, because until the continents grew, there were no shallow seas, and it is these continental shelf seas where most of the frictional drag takes place. Plate motions slowed the spin of the Earth and in doing so increased the number of daylight hours of each rotation. For the blue-green algae dependent on photosynthesis for life, this was a critical catalyst in driving a rapid increase in their productivity. And the product of this productivity? Oxygen. It is because of the evolution of the solid Earth and the ensuing feedbacks between the continents, oceans, and lunar tidal forces that our atmosphere became oxygenated and allowed the evolution of early life.

With a rising abundance of the necessary elements and a planetary environment increasingly favorable for its proliferation, life transitioned from its early single-celled forms into complex multicellular organisms between one and two billion years ago. But the Earth/ocean/atmosphere feedbacks were still far from stable. Processes of mantle overturning, the titans of the Earthly underworld, continued to drive slow and steady change at the surface. More than a billion years ago, plate fragments torn from Nuna once more collided to give rise to a new supercontinent, Rodinia. For three hundred million years, this still poorly defined landmass remained barren, devoid of life. Around eight hundred million years ago, this vast jigsaw puzzle of cratonic fragments felt the inexorable tug of a superplume that heated its base. Thinning and weakening parts of Rodinia, the superplume tore rifts in the continent and allowed new oceans to form. The spreading, newly erupted rock was hot and buoyant, lifting the ocean floor to produce higher sea levels and extensive shallow seas around continental margins. A boon for marine bacterial photosynthesis that accelerated the production of oxygen, the more extensive oceans evaporated more water that in turn rained down on continental land and washed both sediments and the trace elements produced through their chemical weathering back to the seas.

But as the migration of plates continued, slowly reconfiguring the continents into another new pattern of arrangement that by five hundred million years ago would be the supercontinent Gondwana, volcanic arcs at subducting plate margins erupted ash and gas into the evolving atmosphere. Volcanic ash settles quickly from the air, but sulfur dioxide from eruptions forms aerosols—droplets of sulfuric acid suspended in other gases—that block incoming solar radiation. With less sunlight reaching Earth's surface, global cooling quickly ensued. Chemical weathering of new rocks exposed at the surface of the young continents drew carbon dioxide from the atmosphere and exacerbated the cooling. Eventually, the Neoproterozoic climate cooled just enough to trigger widespread snowfall that persisted long enough to form ice sheets across continents mostly distributed around equatorial latitudes. From snow's ability to reflect light arose a positive feedback so effective that the small initial cooling led to ever more cooling until the whole globe was glaciated. Snowball Earth was born.*

Life in the freezer was dramatically curtailed. With ice too thick to allow penetration of sunlight, photosynthesis was no longer viable for much of Earth's early organisms. And with half of the oceans' water locked up in ice, salinity in the remaining ocean doubled, making survival challenging for any but the toughest species. Yet evidence of a mass extinction coincident with the Neoproterozoic snowball Earth remains sparse and ambiguous, suggesting that, somehow, life survived. Warmed by heat from the interior of the Earth, ocean floor hydrothermal vents near oceanic plate spreading centers would have provided safe havens, refugia, for slimy "mats" of bacteria adapted to living in extreme environments. Beneath the monstrous ice sheets, pockets of still-liquid water may have hosted communities of simple life-forms. And above the ice, life perhaps clung to nunataks, mountain peaks protruding above the expansive ice.

Exposed rock was rare, however, and so the capacity for the weathering processes that drew carbon dioxide from the atmosphere was significantly reduced. With less carbon dioxide locked up in surface sediments, atmospheric concentrations of the gas rose, triggering a warming that started to thaw the frigid landscape. The low reflectivity of dark rock and soil, newly exposed by retreating glacier margins, meant more solar energy was absorbed. Dark ocean surfaces, released from their icy restraints, also absorbed heat, and as water from the surface of the Earth evaporated, it formed clouds that trapped longwave radiation—heat energy radiated from the warming world. Collectively this multifaceted positive feedback transformed the global landscape once more and set the stage for the most important and dramatic period of biological evolution in the geological record—the Cambrian explosion.

Around 540 million years ago, during the final stages of the assembly of Gondwana, fossil sequences from around the world suddenly changed.

---

* It has also been proposed that an earlier global glaciation, the Huronian glaciation, occurred during the Archean, from approximately 2.4 to 2.2 billion years ago. Global cooling arose from atmospheric oxygenation that reduced the concentration of methane, a powerful greenhouse gas. It is unclear, however, whether glaciation during this time was global in extent or occurred at different times in different places. Henkes, G. A., Passey, B. H., Grossman, E. L., Shenton, B. J., Yancey, T. E. & Pérez-Huerta, A. Temperature evolution and the oxygen isotope composition of Phanerozoic oceans from carbonate clumped isotope thermometry. *Earth and Planetary Science Letters* 490, 40–50 (2018).

Whereas older strata preserved evidence of primarily simple life-forms—single-celled or small multicellular organisms—within ten or twenty million years, a vast diversity of complex, multicellular life appeared. And most significantly, this was the time that unambiguous evidence of animal life is first found.* As plants and animals diversified and spread, colonizing new lands and new ecological niches, they exerted feedbacks on the environments they inhabited that imposed new controls on the evolving atmosphere. Clay minerals, produced by chemical weathering of continental rocks, became trapped on land by the plants newly spreading across the terrestrial realm. In trapping the clay, those plants helped lock away the carbon dioxide removed from the atmosphere during the clay's formation. At the same time, the proliferation of marine fauna that used silica from the seawater to build their skeletons and shells meant that less of that element was available to form marine clays—a process that releases, rather than stores, carbon dioxide. Combined, the rapid biological changes taking place across marine and terrestrial environments four hundred to five hundred million years ago led to a dramatic change in the global carbon cycle and a drop in atmospheric carbon dioxide concentrations that pulled the Earth's climate out of its former greenhouse state into one much more sensitive to change and, eventually, one that was ideally optimized for the emergence and spread of our own species.

In the hundreds of millions of years between the Cambrian explosion and the rise of Homo sapiens, cyclic changes in volcanic eruptions, climate, sea level, and biodiversity took place, renewing and refreshing the global landscape. Long-term changes in the carbon cycle produced a slow cooling, but one that was repeatedly interrupted by abrupt warming events. Snowball Earth and frozen seas traded places with hot periods whose oceans were warm enough to precipitate limestone from their waters and to accommodate crocodiles at the North Pole. Ocean temperatures reconstructed from carbonate fossils show repeated cycles of abrupt warming followed by slower cooling,[22] echoing a pattern that emerges repeatedly in driven systems operating far from equilibrium. With abundant life on the planet, biodiversity crises also started to appear in the fossil record, each time partially resetting the ecological landscape and promoting the subsequent emergence of new, more complex species.

Plate tectonics enabled the first emergence of life, and through its slow shifting of the continents, the opening and closing of oceans, exposure and

---

* Indirect evidence, such as trace fossils of early wormlike creatures, does exist in older rocks, but the interpretation of such features remains debated.

flooding of land bridges, and the irrepressible growth of mountain ranges, it provided the ongoing stimulus necessary for adaptation and evolutionary change. Positive feedbacks cascaded from one system to another, and some of these responded by storing up energy until a critical threshold was reached. Releasing the stored energy far more rapidly than it had accumulated gave these Earth systems sufficient driving force for accelerated change. Cyclic repetition combined with chance events produced small but significant changes that brought about gradual but purposeful evolution. Through the combination of an oscillating supply of energy from the sun, episodic catastrophes, positive feedbacks, and directed adaptation, the Earth's physical systems slowly but surely entrained and modified the naive and impressionable biological realm that it hosted.

But how did those first experiments in multicellular evolution lead, millions of years later, to biologically complex, technologically advanced, and intellectually curious creatures capable of abstract thought, religion, and science? What were the feedbacks and contingent events that provoked adaptation and locked in the modifications necessary for genetic optimization? And why is it that when we implore our eyes to adjust to the dimness of ancient worlds, to the shadowed palimpsest of times serendipitously secreted in the rocky pages of our past, we find it hard to trace an unbroken path to our origins? What then, were the rules for *Life*?

LIFE

2

Nature! She is ever shaping new forms: what is, has never yet been; what has been, comes not again.

—GOETHE

Of the eight million species on this planet, only one has systematically and deliberately reshaped the natural hierarchy of life. Amounting to only one ten-thousandth of the total mass of biological material (biomass) on Earth, humans have incrementally extirpated countless other species to gain competitive advantage and to control an ever-greater proportion of global resources. Ten thousand years ago, around the closing of the last great ice age, humans were hunting animals that were fifty times smaller than those they hunted one-and-a-half million years ago.[1] And the reason for this reduction in size wasn't climate change. It was because early humans had intentionally hunted the largest available prey to such an extent that species after species went extinct. And then they hunted the next largest. Progressively adapting to reduced prey size no doubt exerted a feedback on our ancestors, pushing them to innovate and refine the strategies and technologies with which they sought, captured, and killed their food. Since that time, during the recent period of climatic stability we call the Holocene Interglacial, human activity has continued its assault on the natural world; recent estimates suggest that three-quarters of life on Earth has been affected by our behavior, at least on land.[2]

And it is in this terrestrial realm where the vast majority of life currently exists—85 percent of global biomass is found on land and nearly all of that consists of plants. Buried in the rocks beneath us are vast colonies of bacteria and archaea—single-celled organisms that lack a cell nucleus—which together make up around 14 percent, leaving only one percent of global biomass in the ocean. In stark contrast to the abundance of simpler life-forms, the diverse species of the animal kingdom make up less than 0.5 percent of total biomass. Of that 0.5 percent, humans constitute only 3 percent. And it is humbling to think that, despite our sense of self-importance, we are, by mass of carbon, outweighed three to one by the combined global mass of viruses.[3] Three-quarters of new infectious diseases arise from deliberate human impacts on the natural environment, either through direct human-animal interactions, intensive farming, bushmeat hunting and the wildlife trade, or the indirect consequences of climate change, environmental degradation, and habitat destruction.[4]

So how did we end up with an ecosystem so ecologically skewed that the animals we farm for food represent more than ten times the mass of actual wildlife? How did one species out of eight million come to dominate life on Earth and to govern the fates of all others? To answer such questions, we first need to investigate what we know of the diversity of life today, how and when it evolved, and what feedbacks have pushed and pulled at life since its earliest, simple origins.

Taking Stock of Life

Published in 1735, the *Systema Naturae* of naturalist Carl Linnaeus set out a hierarchical classification of known species in which they could be grouped and categorized according to shared traits. Organisms were arranged according to a series of ranks: from domains, based on differences in the structure of an organism's cells, to kingdoms, which separate different kinds of multicelled organisms (animal, plant, fungi), progressively down successive levels of segregation to families, within which genetically distinct genera are identified, and species (organisms that are enough alike to produce fertile offspring) within them. Combining genus and species names to give each organism a unique identifier, the Linnaean binomial system proved so remarkably useful that it is still in use today, despite the emergence and popularization of other techniques, such as those that rely more specifically on shared or derived characteristics, an approach known as cladistics.

But although this rigorous classification of biological variability undoubtedly accelerated studies in the natural sciences during subsequent centuries, providing as it did the necessary framework for the genetic and evolutionary ideas of Wallace, Darwin, and Mendel a hundred years later, it was certainly not the first time that great thinkers had sought to bring order to the overwhelming complexity of observed life. In the fourth century BCE, the Greek philosopher Aristotle investigated at length many aspects of the biological world, producing five major works and seven other volumes that collectively expounded upon aspects of biotic life, including animal physiology, locomotion, sleep, and death. Based on his practice of careful observation and rational interpretation (in some ways foreshadowing the reductionist approach of Sir Isaac Newton two millennia later), Aristotle put forward what he called a "theory of form." In this theory, Aristotle grouped organisms into visibly similar kinds—such as birds—and, for each of these, presented the observable features that allowed their unique characterization. Each different "kind" of animal was called a *génos* (from which our current term *genus* is derived). Dividing further, each *génos* was composed of different forms, such that sparrows and eagles represented distinct forms of the same kind of animal. The parallels with the Linnaean binomial system are hard to ignore, even if in some cases the two systems—separated as they were by two thousand years—do not always translate interchangeably.

Aristotle reasonably differentiated creatures with blood from those without and cataloged around five hundred mammals, birds, and fish. In terms of zoological understanding, however, classification is, perhaps, the easy part. More challenging is explaining the observed differences that underpin the groupings. Aristotle organized his interpretations according to five key characteristics: metabolism, temperature regulation, processing of sensory information, mechanics of inheritance, and embryonic growth. He conceived of these as individual but connected components of a single system and put forward the idea that this system defined the soul of an organism. Very much ahead of his time, Aristotle had essentially described the soul as the emergent property of a complex dynamical system, much as we think of consciousness today. He also believed that there was a hierarchy to this complexity, such that the most basic form of life, *bios*, would be the only component present in plants, whereas human beings possessed more advanced animate (*Zoë*) and self-conscious (*Psuchë*) life.*

Augmenting his articulation of these five fundamental characteristics of living things, Aristotle also presented quantitative data that explored the relationships between things. Recognizing certain patterns in his data,

---

* Perhaps coincidentally, the most low-level instructional interface for many modern computers, the code that allows a machine to start up when first powered on, is also known as BIOS—the basic input/output system.

he asserted that larger mammals had fewer offspring and that they lived longer and had more protracted gestation periods—observations that in many cases have proven to be correct. These kinds of broad generalizations are similar to those that underpin the idea of allometry, an approach that seeks to describe how an organism's characteristics or processes relate to one another by way of a scaling law. Despite the rather broad generalizations inherent in such simplifications, remarkable relationships nonetheless arise. Using an allometric approach to investigate the relationship (scaling) between human fertility and energy consumption, researchers at the University of New Mexico uncovered something surprising. Using data from more than a hundred countries and spanning almost thirty years, Melanie Moses and James Brown showed that the average number of offspring born to each female in a population followed a clear negative relationship with their total energy consumption—a finding that confirmed previous studies of metabolic scaling theory as well as the theories of Aristotle much earlier. What was new, however, was that when the data from their human samples were plotted over that from all other mammal species, the relationship held only if nonmetabolic energy sources were also included. That is to say, humans only fit the natural order if we also include the energy we acquire from external sources—such as fossil fuels or renewable energy.[5]

These allometric, or power-law, relationships that appear in the biological realm are phenomena that pervade the physical world. Collectively, these kinds of relationships are considered to reflect a property known as scale invariance, or scale-free behavior, and because these kinds of patterns appear across a vast diversity of systems, they are thought to represent something fundamental about how both natural and human-engineered systems operate. In ecology, some of these relationships have names, such as Bergmann's rule, which dictates the tendency for larger organisms to occur in cooler climates, or the Sheldon spectrum, which shows a remarkable consistency of scaling—spanning twenty-five orders of magnitude*—between the body mass of aquatic lifeforms and their abundance in the ocean. This distribution means that in the ocean today, each group of organisms of a certain order-of-magnitude size contains more or less the same total mass, about one gigaton, regardless of the species from which that total is composed.[6] On land, one of the most well-known examples of scale invariance is the species-area relationship, in which larger areas not only accommodate larger populations, but they also contain more species.

---

* One order of magnitude is equivalent to multiplying something by ten, so twenty-five orders of magnitude is equal to multiplying by ten, twenty-five times over—an immense number.

As a consequence of their greater species diversity, larger habitat areas are also associated with a greater number of connections between individuals and species and a higher number of species carrying out different functions. With increasing size therefore arises functional redundancy—the ability of different species to serve the same functional roles—thereby increasing resilience of the ecosystem as a whole. Functional redundancy is greatest at tropical latitudes and decreases as temperatures drop toward the poles, suggesting that changes in climate might lead to changes in the carrying capacity of an environment.[7] Higher latitude environments, where the conditions change most dramatically and most frequently, favor fast-growing opportunists, species that can make the most of a resource while it lasts. Less changeable environments, those at low latitudes for example, tend to be dominated by "gleaners"—those that can get by on minimal resources. In today's world the distribution of species therefore reflects the waxing and waning of Earth's physical environment over millions of years, each new deviation pushing evolutionary dynamics in a new direction.

Scaling laws such as those described above are more than convenient approximations or entertaining curiosities. That such diverse ecological systems exhibit such regularity in their structure hints at something deeper in the processes that govern not just a single species, but the rich mosaic of life that patterns almost every ecological niche on Earth. Although a multitude of distinct individual mechanisms might produce similar large-scale patterns, feedback processes within these behaviors raise up harmonies from the cladistic choir of multicellular life that offers broad predictability, a baseline for reference when things change. Such is the strength of these allometric laws that systematic departures from them provide good evidence that ecological restructuring is underway.[8] And in the modern ocean, we are now beginning to see that restructuring taking place. Over the last two centuries, the direct effect of fishing and whaling has truncated the Sheldon spectrum, the oceanic size/frequency law. Humans have not only replaced the oceans' top predators, but through our myriad activities have fundamentally changed the way in which energy flows through the system.[9] And from the traces of past lives, captured in geological archives across the globe, we know that a loss of diversity can trigger feedbacks that are devastating. Since the end of the last ice age, the progressive extermination and extinction of ancient fauna has led to a reduced resilience of modern fauna, because it has disrupted the interactions between species and precipitated the collapse of the critical ecological networks that had previously ensured survival. Primates, like us, are now one of the orders of mammals most at risk of extinction, a trend that continues to worsen with each decade.

The First Life

Life, in all its shining exuberance, evolved episodically through eruptions of increasing complexity, enthusiastically leaping between steppingstones of possibility to trace a chaotic path between speciation and extinction. We have described less than a quarter of the eight million or so species on our planet, yet we know they reveal exquisite diversity in form, function, and habitat and must all trace their origins back to a single common ancestor.

What did that "first life" look like? Fossilized remains from three-and-a-half billion years ago in Pilbara, Western Australia, may be the first direct evidence of single-celled creatures. These rocks preserve the mineralized remains of algal mats, built up over time, layer on layer, forming stromatolites (from the Greek *strôma*, meaning "layer," and *lithos*, meaning "rock"). The cyanobacteria that commonly compose algal mats produce energy directly from sunlight, using photosynthetic pigments such as chlorophyll to absorb solar energy and drive chemical reactions. By using light to break down carbon dioxide into its constituent elements, these early colonies of simple life liberated tiny amounts of free oxygen into the shallow ocean. At first this oxygen was hungrily removed by oxidative processes taking place under what was then a weakly reducing atmosphere—one rich in dinitrogen (N_2) and carbon dioxide (CO_2) with small amounts of water (H_2O), methane (CH_4), carbon monoxide (CO), and hydrogen (H_2). Sediments deposited at this time contained grains of iron pyrite (iron chemically combined with sulfur), a uranium-rich ore called uraninite, and siderite, an iron-rich carbonate mineral. Banded iron formations record the transport of dissoved iron to the deep ocean in those first few billion years, but as oxygen levels in shallow marine environments steadily increased from two-and-a-half to two billion years ago, this transport ceased. Progressive oxygenation of the oceans converted ferrous iron into insoluble iron compounds, and the now-famous "redbeds," colored by oxidized iron in the form of hematite, began to appear in the geological record.

The Great Oxidation Event (GOE), as it is now known, began a billion years after those first Australian stromatolites locked microscopic life into their stratified form and played out over four hundred million years. And it represents one of the most important feedback to have ever taken place on Earth. Before the GOE, all life was anaerobic, existing without oxygen and relying on carbon, nitrogen, hydrogen, and sulfur from their environment to supply the compounds necessary for energy production. But anaerobic respiration is less efficient than its aerobic cousin, releasing less energy and keeping the pace of cellular life slow. So although life initially took hold when oxygen was scarce, it was the increasing release of free oxygen into the world that enabled life to thrive. At first, the carbon and oxygen molecules freed from ocean water by photosynthesis readily recombined with each other. Only when other processes sequestered the carbon, burying it in sediments and preventing its oxidation, was oxygen finally liberated from the ocean to transform the methane-rich atmosphere.

Nearly all life today needs oxygen to survive. But two billion years ago, rising atmospheric oxygen levels were as much of a threat to early life as rising carbon dioxide is for us now. For species that had evolved under a reducing atmosphere, oxygen was poisonous. Yet the steady production of oxygen—a waste product of photosynthetic bacteria colonizing the global oceans—eventually polluted the Earth's atmosphere to such an extent that it led to widespread extinction of early anaerobes and a collapse of primary productivity that forever changed our evolutionary trajectory.

For the change in atmospheric composition didn't just kill off the anaerobes, it presented a new opportunity for those bacteria whose metabolic functioning could use the newly available oxygen and the greater source of energy it represented. Harnessing that energy was the challenge. Today, 99.5 percent of described species on Earth are eukaryotes, organisms whose cells contain a nucleus.[10] Eukaryotic cells are relatively capacious, with complex internal structures that allow different cellular functions to be separated. The nucleus houses genetic material needed for reproduction, whereas an organelle called the endoplasmic reticulum is responsible for protein folding and assembly. But it is another organelle, the mitochondrion, that is most important in understanding the impact of atmospheric evolution on that taking place in the oceans beneath. Mitochondria are double-walled cellular subunits that generate energy from oxygen-driven (aerobic) respiration. They have their own genome and are so like bacteria that one possibility for their origin is that they started out as independent prokaryotes (single-celled organisms lacking a nucleus) that became incorporated into other forms of early life, such as archaea. Archaea are single-celled prokaryotes, like bacteria, but are distinct enough that they constitute an entirely separate domain (the highest level of the Linnaean-based taxonomic classification). One hypothesis for the origin of complex life, therefore, is that an archaeon, attempting to increase the surface area of its exterior membrane to gain greater access to nutrients, developed protrusions that eventually wrapped around adjacent bacteria. A symbiotic feedback ensued in which the benefits of mutualism allowed for the proliferation of an entirely new domain, the eukaryotes.

Trapping separate components within a single structure allowed multiple functions to be carried out within a single eukaryote cell, and the advantages that this new configuration conferred initiated a positive developmental feedback that allowed the first eukaryotes to become the ancestors of all complex life. Early eukaryote organisms, such as protozoa, grazed on bacteria in the ocean. And the bacteria fed on dissolved organic matter freed up by phytoplankton, which in turn were fueled by nutrients released by the eukaryotes. This so-called microbial loop is a three-part cycle in which protozoa, phytoplankton, and bacteria are connected through a continuous cycle in which they each benefit from nutrients released from or absorbed by one of the others. In 2004, the physicist Fritjof Capra, perhaps most well-known for the groundbreaking 1975 book *Tao of Physics*,[*] developed the idea of functional cognition in an evolving system.[11] He argued that any system absorbing an external source of energy does so in a way that is governed by its physical structure but that this structure is itself shaped by the energy it is absorbing. Capra proposed a three-part cycle, just like the microbial loop. First, the constructional form of an object (or organism) influences the way that energy is absorbed. Then, processes within the system control how that energy is either used internally or dissipated to the external environment. Finally, connections within the system are modified (structural adaptations) to regulate the flow of the absorbed energy and allow it to make better use of the supply available. Repeatedly cycling through this loop leads to incremental refinement of the system and a sequence of gradual adaptation that improves the system's efficiency. This is the "cognitive" part of the cycle—the system essentially "learns" how to operate more effectively. This kind of feedback-driven adaptation within a multicomponent biological system promotes evolutionary change because it allows the system to make the best use of the energy available in its environment. And by becoming more efficient, an organism becomes more resilient to change or can more effectively exploit new habitats.

In the cognitive cycle of the microbial loop, tentatively established in the sun-drenched shallow waters of the early oceans two billion years ago, feedbacks strengthened the system with each new iteration. And as this dynamically adapting system became increasingly well-established, it grew, benefiting from structural and chemical adaptations that allowed more efficient cycling of nutrients and the more efficient use of energy that arose from the still-new process of aerobic respiration.

The evolutionary leap from single to multicelled organisms, however, most likely required more than just the efficient use of energy. Yet the

---

* In the *Tao of Physics* Capra illuminated the common ground between the quantum physical world, characterized by probabilistic measurement and associated uncertainty, and the wisdom of ancient eastern philosophy known as the Tao.

clarity of our fossilized windows into this distant world is disappointingly subpar, leaving us, the observers, to try and reconstruct from shadows what must have been truth. In doing so, we find perhaps three likely candidates to explain the occurrence of multicellularity. One idea is that single-celled organisms agglomerated, working together as a single entity like slime molds do today. Another suggests that multiple cells may have arisen from a single cell in which the nucleus somehow split in two, over and over, each time creating partitions—cell walls—between the divided components to eventually produce a creature consisting of a group of connected cells. A third hypothesis is that multicellularity could occur when a single cell divides if the daughter cells fail to separate. Whatever the truth may be—either one, two, or all of these—we know that multicellularity in eukaryotes has evolved at least two dozen times.

Whereas single-celled organisms were limited in size by their ability to absorb and transport nutrients through their walls, creatures with multiple cells could grow increasingly large without such constraint. Not only that, but an organism composed of multiple cells could live longer, because death of individual cells posed considerably less threat to overall viability. The competitive advantages afforded by multicellular structures became a positive feedback that accelerated the rise of complexity, provisioning larger and longer-lived creatures with enhanced opportunities for cellular differentiation within a single organism. From their uncertain and perhaps serendipitous origins, complex life-forms had by now acquired nearly all of the necessary capabilities and resources needed for their propitious explosion into the eight million species we estimate to exist today. The biotic world was still tied to the vagaries of geophysical phenomena but was increasingly becoming a directing force in its own right, manipulating the physical environment in ways that favored ecological continuance. But staging of this global coup took time, from the first emergence of eukaryotes one-and-a-half to two billion years ago to the Cambrian explosion more than a billion years later. Why the delay?

If the Great Oxidation Event of two-and-a-half billion years ago had set the stage for the emergence of eukaryotes, then it was to be another such event—the Neoproterozoic oxygenation event of seven hundred million years ago—that oxygenated the atmosphere and catalyzed the rapid expansion of complex life. Although the GOE had transformed the global ocean into one that could support faster and more energetic chemical reactions, it had done little to alter atmospheric chemistry. The problem was that the amount of oxygen escaping from the ocean was still so insignificant that it was rapidly used up in oxidation processes at the land surface.

So it wasn't that oxygen wasn't being produced, it was that the feedbacks between the biologic, oceanic, and terrestrial realms tended to be self-limiting and allowed only incremental modification of the coupled system. To drive accelerating change of the type characterizing the Cambrian explosion 541 million years ago, something extra was needed. As far as we know, there was no external source for the vast amount of free oxygen that would have been necessary to drive rapid atmospheric change as recorded in paleoenvironmental archives. But if the source wasn't external, it must have come from whatever geophysical and biological processes existed at that time. And that means that the abrupt shift in oxygenation that took place must have been an emergent phenomenon of the coupled earth-biosphere system. But what triggered it? A complex system, composed of parts that each dynamically interact with one another, can undergo rapid changes (phase transitions or regime shifts) if one part of the system is sufficiently disturbed that it can no longer limit change in another. With one component unregulated, it drives the system increasingly further from its prior point of balance, a positive feedback that eventually culminates in a substantively different state. And according to recent theories of the Neoproterozoic oxygenation event, that's precisely what happened.

Rather than a single trigger producing pulses of oxygenation throughout the 4.53 billion years of Earth's history, such as a rapid expansion of photosynthesizing plankton, it now seems likely that these pulses arose episodically because of feedbacks between the long-term biogeochemical cycles of carbon, oxygen, and phosphorus. These elements pushed and pulled one another in a slow-motion *mêlée à trois*, each planetary pugilist beating down another only to be beaten down itself by the third. Phosphorus, for example, controls ocean productivity—the growth or collapse of phytoplankton species—which in turn controls the production of oxygen through photosynthesis. When these phytoplankton die and decompose, however, the carbon, phosphorus, and nitrogen they contain combines with dissolved oxygen to form carbonates, phosphates, and nitrates. Settling to the ocean floor, these new compounds lock away in deep sediments some of the oxygen that was previously available for aerobic life. Above the ocean, global temperature and pressure gradients drive winds that mix the ocean beneath, transporting nutrients both vertically and horizontally and in doing so continually altering the conditions for life at the surface. Phytoplankton blooms, responding to the swirling eddies of nutrient supply, thus dictate the balance of oxygen and carbon in the surface ocean and in the air above. Depending on whether the ocean soaks up or releases carbon dioxide, it controls climatic feedbacks, regulates aerobic biological activity on land, and controls surface weathering that supplies back to the ocean many of the nutrients on which marine organisms depend. Geographic variations in the distribution of these nutrients, as well as their interdependent fluctuations through time, thus conspired to produce a stepped change in the availability of oxygen for life on Earth. And by 541 million years ago, an ecological tipping point had been reached.

The Explosion of Life

In the mid-nineteenth century geologists found themselves confounded by a worrisome problem. In rock strata older than about 540 million years, there was a distinct lack of fossils. Yet somehow, a large and diverse population of species appeared in the rocks of the Cambrian period immediately above this layer. It was almost as if they had come out of nowhere and instantaneously populated the globe. At a time when Charles Darwin's theory of natural selection was gaining increasing acceptance, this geological ambiguity was immensely troubling. If Darwin was correct in his proposition that each species had evolved slowly and incrementally from a similar but somehow different ancestor, then where were the ancestors of the abundant Cambrian fossils? Were the fossils missing? Or were they never there? To this conundrum, the by-then-famous naturalist had "no satisfactory answer."[12]

Over the century and a half that followed, the view of this remarkable transition became clearer. As we saw earlier, simple unicellular life existed before the "Cambrian explosion"—stromatolites and phytoplankton dating back billions, not millions, of years. And the ascendence of the eukaryotes, catalyzed by rising oxygen saturation in the ocean, then marked the beginnings of multicellular life a billion or more years later. But it was only around seven hundred million years ago that the second major chemical transformation of the Earth, in the form of the Neoproterozoic oxygenation event, brought free oxygen to the atmosphere. There is evidence of complex marine life emerging before the Cambrian explosion, during a period known as the Ediacaran, but the expansion of life from ocean to land occurred a hundred million years later. And once it started, it took place blisteringly fast—in geological terms at least. Over a period of just a few tens of millions of years, the first primitive shallow-water plants, aided by fungi, established themselves on the shores of new and uninhabited lands.

Evidence of the transformation that followed abounds. In 1909, Charles Walcott, then secretary of the Smithsonian Institute in Washington, D.C., discovered fossil beds in British Columbia, Canada, whose state of preservation astounded the world. Perhaps because of brine seeps in the ocean floor, where heavily mineralized waters from tectonic faults reached the seabed, the 508-million-year-old creatures entombed in the Burgess Shale formation were archived so precisely and with such little degradation that they truly represented a window into a world long lost. The unusual conditions of their burial allowed even soft parts of these first complex life-forms to be mineralized, recording an astonishing completeness of anatomy. Shortly after Walcott's discovery in Canada, in the years before World War I, a botanical counterpart to the animal-rich Burgess Shale was unearthed in Scotland by geologist William Mackie. Though younger by one hundred million years, these relics once again exhibited exceptional states of preservation. Delicately mineralized in the waters of volcanic springs, the plant fossils of the Rhynie Chert were petrified so quickly that individual cells could still be seen. Fungal filaments—hyphae—were also identified, growing into those early plants and confirming the vital symbiotic role they played in the establishment and ecological success of their hosts, just as they do today.

With plants now expanding across the pristine terrestrial realm, first invertebrates, then vertebrates, followed quickly thereafter. The freedom of these new and untested lands allowed rapid proliferation and the diversification of species. Exploiting niches never before occupied, the genetic foundation of these early creatures had almost free rein to express itself unimpeded, daringly provoking the naive world. Mathematically we might describe the pioneer spirit of these exploratory genomes as one that "allowed large deviation trajectories" that led to "quasi-equilibrium states."[13] Basically, the field was wide open, and there were many directions to run. If enough genetic code is allowed enough freedom to replicate with enough frequency, ecological diversity is almost guaranteed. And it is just such a scenario that appears to best explain this explosion of life. All the possibilities were there, laid out like keys on a piano waiting to be played. The performance, however, still needed its maestro.

The constant jostling between the chemical components of the early ocean had given rise to pulses of oxygenation that bubbled forth over hundreds of millions of years whenever the balance of nutrients tipped just a little further in a particularly favorable direction. Now, on land, it was feedbacks between the atmosphere, the Earth, and the newly emerging life upon it that slowly but determinedly set the rules. The genetic patterns of the first complex creatures therefore arose like a sequence of notes from an amateur musician, as effervescent possibilities but no more;

they were not retained. But with practice came refinement, and remembering, and selective retention, such that our ecological musician began to keep only those sequences that were most harmonious, that best fitted the symphonic landscape afforded by the wider environment. And as each of the assembling players found its niche, and the bounds within which it could contribute to the genetic pool, the emerging fanfare rapidly burst forth in a triumphal exuberance of life, converging on the evolutionary melody we still sing today.

Biodiversity and Its Crises

Viewing speciation—the emergence of new species—as a phase transition driven by random (or stochastic) events aligns perfectly with the previous evidence we have seen in other dynamical systems, from the initiation and progressive organization of plate tectonics to the incremental recombination of six essential elements that eventually produced all the nutrients necessary to catalyze and support early life. And one of the attractive aspects of such a view is that the emergence of rapid changes and seemingly isolated yet abrupt events requires nothing more than the existence of feedbacks. Specifically, it was the pressures arising from competition, predation, and environmental change that cajoled and coerced each new species to either evolve or be supplanted by a better version more likely to be retained. In this way, the "blind evolutionary process"[14] gives rise to long-lasting genetic selection that favors organisms whose structure and function is aligned with the external environment in ways that maximize the organism's survivability and reproduction. Genetic diversification is typically faster in groups of organisms (clades) that are still close to their point of departure from a common ancestor compared to those that have continued along an evolutionary trajectory for somewhat longer. This may be because each successive genetic change locks in a particular way of being, such that a path dependency arises; the options for genetic deviation decrease through time because changing might be risky for an organism that has already proved its evolutionary viability.

Eventually, of course, all species must go extinct, and in the fossil record a pattern of long-rolling genetic turnover emerges. Given the diversity of biological form and genetics observed in this record and the prevalence of abrupt environmental changes, geophysical upheavals, extraterrestrial bombardments, and so on, it seems likely that the roiling cauldron of evolutionary change should be one exemplified by chaotic change and unpredictable events. Such cataclysms did of course take place, but against all the odds, evolution seems to have proceeded in fairly regular cycles of diversity that waxed and waned through time. When the early naturalists began piecing together the fragmentary evolutionary jigsaw that was the geological fossil record, they relied on rock strata between which there were large and obvious changes in flora and fauna, just as we saw earlier at the beginning of the Cambrian period. These abrupt transitions came to define the geological time line and allowed creatures of particular periods to be considered together, as similar in form and age. As a consequence of this approach, periods when life was abruptly replaced by newer, markedly different forms were identified as "mass extinctions," after which "creative evolutionary radiations" represented the subsequent recovery of the biosphere. The concept of mass extinction is an emotive and engaging one, but an increasing number of studies are bringing new perspectives to the idea that the boom-bust cyclicities of life on Earth were isolated and dramatic events that brought life to its knees in an instant, only to rise, quiveringly, to regain its composure millions of years later.

Some of this new research reveals that species tended to originate, proliferate, and die out according to cycles that each vary in length but average approximately twenty to thirty million years.[15] In some cases, this length of time aligns with the spacing between the so-called big five mass extinctions, for example between the end-Ordovician extinction at around 444 million years and the Silurian-Devonian extinction at 419 million years. But the spacing between other mass extinctions is far longer, with the next youngest being the end-Permian extinction (252 million years ago, or Ma), end-Triassic (201 Ma), and the famous-because-of-the-dinosaurs end-Cretaceous mass extinction at 66 million years ago. The reason for this irregularity is that, when it comes to evolution and species turnover, mass extinctions represent only half the story. The other half is that of origination. For example, if the rates at which new species originate and existing species go extinct are similar, the fossil record documents a story of smooth and incremental evolutionary change; that is, no mass extinctions occur despite continual and fairly regular ecological turnover. Yet if *either* of these two rates increases over the other, the chronicles will tell a tale of species-specific extinction. Indeed, many of the apparent extinction events identified since complex life has existed on the planet have either differentially affected particular species or animal groups or were restricted to land or to the ocean. And preceding some of these "extinction events" were periods when significant diversification was taking place. Other extinction events were exacerbated by a slow and steady decline in diversity as origination rates fell, rather than being caused solely by catastrophic environmental impacts that obliterated animals in their prime. So depending on where we look and what we measure, we might see a different story.

Recently, a team led by British scientist Jennifer Hoyal Cuthill coined the term "evolutionary decay" as a new way to frame this idea of time-evolving changes in the balance of speciation and extinction.[16] Fundamental to this concept is the idea that species can disappear from the fossil record in the two ways we just saw—they might go extinct, but they also might be crowded out by a rapid proliferation of other species. The "evolutionary decay clock"[17] is a way to measure the pacing with which the relative number of coexisting species rises and falls over time. Considering the last five hundred million years of complex life, this new approach shows that when considering the entire assemblage of species that exist at any one time, few of them overlap with one another for more than twenty to thirty million years. This average hides considerable variability, of course, and the typical persistence lifetime of any single species during the last half-billion years may have been anywhere between one and fifty million years. Some studies even identify "ephemeral species" that persisted for less than a million years.[18] But the major periods of evolutionary disruption—marked by discontinuities in the fossil record—tend, therefore, to reflect times when long-lived species were at a minimum, perhaps because of an environmental catastrophe, and when young or new species were perhaps more prevalent as a proportion of the whole biota, although they could have still been few in absolute number. Either way, one of the most intriguing findings from this research is that these cycles in genetic diversity are far from random. Instead, a regular pattern emerges in which the proportion of overlapping species increases steadily over tens of millions of years, only to then drop much more quickly, tracing out the same kind of sawtooth pattern we encountered previously in the rainfall patterns that dominated the extreme hothouse climate of the early Earth. And as we discover in later chapters, this asymmetric pattern of global speciation and extinction is just like the behavior we see in the electrical activity of our hearts or in climate fluctuations spanning hundreds of thousands of years.

In identifying this common trait, we gain predictability through transferability. Knowing that asymmetric or sawtooth oscillators arise in systems where positive feedbacks increasingly drives a system further from equilibrium until it is no longer tenable and thus rapidly collapses, we can apply a new lens to the problem of evolution. One of the ideas surrounding the concept of mass extinctions is that these dramatic events somehow wiped clean the genetic slate and facilitated rapid subsequent evolution of new species, perhaps because ecological niches were left vacant for novel species to then adopt. This may be true to a certain extent, but some evidence suggests otherwise. The largest radiations (periods of species diversification) didn't necessarily follow periods of correspondingly large reduction in genetic diversity. Instead, the periods of most rapid genetic diversification occurred "when life exploited new realms of opportunity."[19] The Cambrian explosion, for example, occurred because feedbacks between geophysical and ecological systems enabled life to escape the oceanic realm and expand onto land, the terrestrial realm. Magnitude of loss, therefore, does not necessarily dictate magnitude of regrowth. Furthermore, the *rate* at which life recovered after each evolutionary bottleneck appears to have been similar, regardless of the scale of the bottleneck itself. Recovery can only take place so fast—there seems to be an intrinsic speed limit on how quickly the ecological system can reboot—and this recovery is most likely controlled not by what went before, but by the dynamics of the genetic diversification process itself.[20] Species don't just occupy ecological niches, they themselves are also niches for other predatory or symbiotic species that might depend on or benefit from them. Radiation therefore proceeds slowly but accelerates quickly as species diversify and give rise to ever more opportunities for life and new evolutionary pathways. Diversification creates the niche space that enables further diversification—a positive feedback. If the emergence of novel ecological niches fails to keep up with the rate of diversification, this self-sustaining process might eventually reach a steady state of turnover. The slow start, rapid acceleration, and subsequent stabilization once again traces out a familiar pattern, that of a phase transition or regime shift.

There were, of course, periods when significant diversity crises did occur. Of these, the end-Permian event of 252 million years ago is considered the most substantial, with more than 70 percent of species thought to have gone extinct, maybe in as little as a few tens of thousands of years. This episode affected both terrestrial and oceanic realms and was associated with large-scale changes in the global carbon cycle. Most likely initiated by one or two closely spaced pulses of magmatism in what is now Siberia, the release of volcanic gases caused atmospheric carbon dioxide to rise rapidly and led to an atmospheric greenhouse effect that not only raised air temperatures by 10 degrees Celsius but also warmed, acidified, and deoxygenated the global oceans.[21] Though these impacts may have been felt at different times, perhaps over tens or even hundreds of thousands of years as the slower-responding ocean absorbed heat and carbon dioxide from the atmosphere above, a multitude of environmental changes ensued that set in motion a cascade of feedbacks. The warmer atmosphere drove enhanced hydrological cycling, changing not only rainfall patterns but also its intensity, in turn accelerating erosion and riverine transport of sediments eroded from strata that were now weathering faster in the more humid atmosphere. Impacts on plant life would cascade into the fauna that relied upon them, with the worst-hit terrestrial species being the herbivores who couldn't adapt to the abrupt floral changes to their habitats. Collectively these feedbacks gave rise to a suite of kill mechanisms that affected nearly all life on Earth.

Although devastating, the sequence of events that played out during the end-Permian event illustrates that it is the feedbacks between the coupled components of the Earth system that decide the fate of extant life and the scale of impact of seemingly random external catastrophes. One theory, for example, proposes that the Siberian eruption ignited coal beds that released additional carbon dioxide into the atmosphere, further exacerbating the climatic warming and, by liberating toxic and radioactive elements from the coal, presented additional challenges to an ecosystem already struggling for survival.[22] As we have seen previously, the story that emerges is not a simple one in which "mass extinction" can be attributed to a single mechanism; instead is revealed a complex system of diverse but coupled components, or oscillators, each driven further from equilibrium both by incremental forcers and by stochastic events. Irreversibility in these systems locks in incremental change so that each one of these components independently reaches a critical state. All that is then needed is a small perturbation that, through the global network of connected components, drives a sweeping cascade of self-reinforcing feedbacks that brings about widespread reorganization and the establishment of a completely new ecological state.

Returning to our musical analogy once more, we are presented with a lingering problem. Our genetic musicians, the diverse species of life on Earth, evolved increasingly rapidly once allowed to do so, complementing and harmonizing with one another as they each first established then cemented their niche in the ecological orchestra. And as with any symphony, the melody was carried by different players at different times, each lending its own interpretation, emphasis, and dynamic to the organic serenade. Yet our orchestra is lacking its conductor, the guiding hand that holds back or brings forth those sections requiring either subtlety or dominance, the hand that watches over the collective, instills unity, and enables the emergence of mellifluous beauty. In 1759, the economist Adam Smith envisaged an "invisible hand" that would distribute to individuals an equal measure of the resources required to achieve collective betterment of society and the "multiplication of the species."[23] Where, then, in the explosions and bottlenecks of speciation and extinction should we look for evidence of this conductor, this invisible hand?

When, in 1859, Charles Darwin published *On the Origin of Species by Means of Natural Selection*, the idea that the creatures around them had somehow come into being by chance, followed by a process of slow and gradual change, was too much of a challenge to their belief in divine creation for some people to accept. Particularly prickly was the suggestion, developed in his 1871 work *On the Descent of Man and Selection in Relation to Sex*, that humans, revered as the epitome of intelligent life and quite above the savagery of the natural world, could have in fact evolved from the same recent ancestor as the African great apes. So difficult to accept was the idea that humans might be part of the same natural order of things as all other forms of life, that even Alfred Wallace, who communicated and even published with Darwin on the idea of natural selection, remained convinced that humans could only have been a result of divine creation.

Scientific progress over the century and a half since that time—much of it following Newtonian methods of observation and interpretation—has now afforded us a wealth of evidence whose diversity and accuracy irrefutably support the basic Darwinian concept of evolution by natural selection. But far from revealing a process of smooth and gradual change, these data have illuminated just how episodic the leaps and spurts of genetic diversification have been. We saw earlier that if the overlapping lifetimes of entire assemblages of species are averaged in the appropriate way, we can identify time-varying changes in species longevity that resemble a sawtooth oscillator, fluctuations in biodiversity that rise and fall every twenty to thirty million years. But the geological looking glass through which we must peer to deduce such generalizations is clouded and cracked, and where fragments are missing we must infer what was lost. Despite all the post-Darwinian advances we are still, to some extent, in the dark.

And the murkiness with which we contend invites, perhaps even demands, speculation. In relation to nature, the poet Goethe* suggested that "Whoso cannot see her everywhere, sees her nowhere rightly," but what is the "right" way to see nature, to reconstruct lost lives and the forces that guided them? Statistical analysis, in its myriad forms, becomes an important tool for such a quest. And on the basis of many such analyses, evidence emerges of regularities in the occurrences of "mass extinctions,"

---

* Johann Wolfgang von Goethe (1749–1832) was a German polymath widely considered one of the brightest minds to have ever lived and credited with an estimated intelligence quotient of 180.

or "diversity crises." Earlier we saw how geological and geophysical studies have repeatedly identified a frequency approximating thirty million years in records that represent global sea-level oscillations, marine and terrestrial faunal diversity, volcanic activity, spreading rates of the seafloor, or the oxygenation state of the waters above. Are these players the conductor our orchestra is missing, the invisible hand that sets the volume and tempo of change through the biological feedbacks that emerge as they interact with one another? Or are they, too, just part of the orchestra, guided by a far bigger, more distant conductor?

Whether the periodic rise and fall of global biodiversity is evenly paced or temporally haphazard, the evidence at our disposal certainly supports the notion that the invisible hand at work conducting and directing the magnum opus of which we are part is one that relies on coordination between its players and feedbacks that occasionally propel our gospel rapidly forward. But it is one that also does not shy away from intermittent catastrophe. As with all the complex dynamical systems described in other chapters, life on Earth appears to have bounced repeatedly between periods of relative stability and episodes of cataclysmic change. At the end of the Permian, 252 million years ago, the geologically rapid release of carbon dioxide from Siberian flood basalt eruptions triggered a chain reaction in Earth's physical systems that made the global environment insufferably hostile to most of the life that had evolved to thrive under quite different conditions. Since the mid-eighteenth century and increasingly so from the mid-nineteenth century, human activities have also polluted our atmosphere with carbon dioxide, and at a rate that continues to increase year after year. From the end-Permian and other periods of rapid warming in the geological past, we know the sequence of events that follows from elevated carbon dioxide: global warming of our atmosphere and ocean, acidification and deoxygenation of seawater, and a more turbulent climate whose rainfall regimes shift the zones of habitability for many of the plants on which all animals depend. From the end-Permian and other diversity crises, we know that environmental feedbacks, if sufficiently severe to drive widespread extinction or reduced speciation, will trigger a loss of biodiversity that will take five to ten million years to recover.

Is There a Sixth "Mass Extinction"?

The earliest evidence of bipedalism dates to somewhere between six and seven million years ago, ascribed to upright-walking hominids of the *Sahelanthropus* genus.[24] As we see in chapter 4, by three-and-a-half million years ago, bipedalism in the younger *Australopithecus* genus was refined, habitual, and dominant over four-limbed locomotion. But it was another million or so years before speciation led to the branching of the genus *Homo*, to which we belong. And it is thought that *Homo* was the first genus to evolve cultural traits that we might recognize today, such as the creation of hunter-gatherer societies and the use of fire. Species of *Homo* migrated out of Africa into Europe and Asia, taking with them skills and cultural practices from their African origins. And so it was that around two million years ago, humans began to exert an increasing pressure on the natural environments into which they spread. Climate was also changing throughout this period, oscillating intermittently between ice age conditions that allowed vast ice sheets to expand across northern continents and warmer interglacials that allowed plant, animal, and human expansions into previously barren landscapes. The waxing and waning of the global climate led to gradual environmental changes over thousands or even tens of thousands of years, yet the challenges exerted by these natural variations were rarely substantial enough to drive widespread species loss on their own.[25] The increasing presence and impacts of humans, have, however, pushed species to extinction in nearly all habitats on Earth, particularly so within the last few centuries. Fossil evidence shows that until relatively recently, many food webs were much richer than they are today, with longer connective chains and more interactions among species. As the rate of species extinction accelerates, those that cannot adapt quickly enough to changes in their crumbling support systems—their habitat, climatic niche, or food web dependencies, for example—will also disappear. In the absence of migrations into new, more tolerable refuges, many animal species will not survive future warming unless their rate of adaptation is a thousand times faster than rates seen in the fossil record.[26]

To understand the scale and fragility of the current ecological system, we need to understand the roles of mutualism and synchrony—that is, the way that species depend on each other both directly and indirectly. Clearly if a species depends on another that goes extinct, they too will go extinct unless they are adaptable and find an alternative species that can play the same functional role as that which was lost. The size of the web of mutual dependencies therefore controls the magnitude of the potential extinction cascade that might take place.[27] Coextinction accelerates where ecosystems are composed of mutually dependent species, but in many ecosystems there are multiple interaction types, with predation and exploitation existing alongside mutualism. Inevitably there will always be winners when there are losers, and this gives rise to asynchrony between species.

In the 1920s, the first mathematical representations of species interactions began to appear and gave rise to a set of simple equations now commonly referred to as the predator-prey relationship—something we return to in chapter 3. A predator-prey system represents a number of components that interact with each other, in this case competitively, such that their fates are intimately entwined with each other. A rising population of prey, for example, may allow an explosion in the predator group. But with increased predation, the prey numbers then dwindle, followed quickly by those that depended upon them. With predators gone, the prey can once more recover, and so the cycle continues, tracing out a limit cycle of stable but periodic oscillation. But it is not just other species that an organism affects. Commonly, many species act as ecosystem engineers, adapting and shaping their environment in ways that persist beyond an individual's life span. Arguably humans have been the most obvious engineers in this respect, but in the nonhuman natural world, engineering plays an important role in ecosystem stability. We saw in chapter 1 how photosynthetic bacteria radically altered the chemistry of Earth's early atmosphere and allowed complex life to evolve. Beavers build dams that alter river flows in ways that may both help and hinder other riverine species, whereas other organisms such as trees build new habitats for other species as they themselves grow. And with engineering it seems that although an ecosystem with only a few engineers may be fragile, the increasing redundancy that comes with an increasing number of engineer species actually tends to stabilize an ecosystem by providing many more opportunities for colonization and a reduced dependency on any one engineered state.

Species richness, diversity of interaction types, asynchronous dependence, and duplication or redundancy of function are all aspects of an ecosystem that might help survival in the face of environmental change. But unless an ecosystem can adapt its composition, structure, or functional relationships fast enough to keep pace with external pressures, it will ultimately be destined for collapse. Half of all the plant species on Earth rely on animals to spread their seeds.[28] As animal species are lost through hunting, habitat destruction, or extermination on the basis of competitive exclusion, their functional benefit to dependent plant species is also lost. Without seed dispersal, plants that under today's changing climate need to migrate hundreds of meters to tens of kilometers each year to track their climatic niche will not survive; only those that are able to evolve in situ will cope with the current rate of environmental change.

Examples of rapid evolutionary adaptation do exist, however. Intensive ivory poaching in parts of Africa that decimated some elephant populations by as much as 90 percent in only twenty years became such a strong selection pressure that it has since led to the emergence of tusklessness in the survivors.[29] In the human realm, how well placed are we to adapt to our rapidly changing global environment? As our climate warms, shifting rainfall patterns create havoc with arable crops that rely on a stable, or at least predictable, weather cycle. In developing countries, the pressure exerted by repeated dry years has led to an increase in deforestation as land is cleared for crops in an attempt to compensate for declining yields.[30] In the boreal summer of 2021, it was too much rather than too little rainfall that became the problem, both in Europe and China, where flooding displaced more than a million people, left hundreds of thousands without power, and caused property damages amounting to tens of millions of dollars.

Today we rely on irrigation for 40 percent of global food production, and 20 percent of global water use comes from groundwater.[31] Currently, half of the world's population lives in places where groundwater aquifers are being emptied faster than they are being refilled, leading to the proposition that in some places, the time of "peak water" may be rapidly approaching. Unless annual crop production can be balanced with aquifer recharge, a global food supply crisis will be inevitable. But perhaps more concerning than any of these is the discovery that our species will increasingly struggle with global warming not just because of challenges that are physical, but also because of those that are mental. Evidence from sixty countries spanning the years 1979 to 2016 shows that the increasing frequency of heat and humidity extremes is significantly correlated with suicide.[32] Although the triggers for any individual case are difficult, if not impossible, to determine precisely, the data suggest that perhaps increased physiological stress leads to increased mental stress, or that medications used for treating mental illness may decrease the body's ability to thermoregulate, or even that the disruption to a person's social networks and livelihoods caused by heat waves may be a factor in increased anxiety.

In light of the mounting and irrefutable evidence of global climate change, many scientists are now calling for drastic action, even advocating for civil disobedience.[33] One of the concerns among those who study the evidence is that the future into which we are headed might be far more hazardous, more dangerous, than we tend to think. Much of this concern stems from one of the key aspects of many of the feedbacks we have explored so far—that of a time delay between any initial driver and the subsequent response. Just as the faunal landscape of our current global ecosystem is the delayed expression of millennia of hunting and habitat destruction that has systematically altered the natural distribution of species within differing size classes—like the Sheldon spectrum seen earlier—so too it is in the floral realm. In the last three to five thousand years, the rate of vegetation change has accelerated to rates not seen in at least the last eighteen thousand years—this despite the global-scale changes in climate that took place at the end of the last ice age, culminating around ten thousand years ago. The recent and increasingly rapid changes in biodiversity portend a grim future because the impact of these shifts are not yet fully felt. Instead, it will be our children's and perhaps grandchildren's generations who will face the starkest socioeconomic consequences of the current biodiversity crisis.

But how do we gauge the severity of the situation in which we find ourselves? How is it possible to measure and monitor the rate of change of the eight million species on Earth when we have yet to even describe 75 percent of them? And when cyclic patterns of extinction and speciation naturally conspire to muddy the fossil record of long-term evolutionary trends, how might we piece together the fragmented jigsaw of life that we know today and meaningfully construct what is real? Certainly the challenges are immense and the scientific community divided. To make statements of any substance, we must generalize and consider the largest view of the world's ecological systems, but in doing so we miss detail and nuance. Balancing these two becomes a matter of subjectivity, and not everyone agrees on the balance. Predicting the timing of abrupt ecological disruption from future climate change, Christopher Trisos and colleagues[34] foresaw a situation in which continued high emissions of greenhouse gases would initiate impacts as soon as 2030. By mid-century under the same scenario, they envisaged that sudden shifts in ecological assemblages would take place that would put constituent species into climatic contexts that might not be survivable. And because the ecological assemblage includes balanced mutualisms as well as predation and exploitation, all coexisting within a narrow climatic window, climate-driven disruption to one species will be felt by all, with cascading feedbacks rippling quickly through the entire population. But others argue that such broad-scale analysis conjoins species whose habitats may be proximal in terms of horizontal distance, but fundamentally different if they occupy different climatic zones in steep mountain areas, for example. Conflating all species into a single grid cell of a global model underrepresents the significance of local-scale ecological complexity. Regional refugia or habitats that are physiographically separated may aid the survivability of certain species within an area, delaying or even preventing large-scale ecological collapse.[35] So it is the connectivity among species that mediates the feedbacks driven by external environmental change. Without connectivity, there are no cascades, and changes are local and discrete—a theme that we pick up again in chapter 5. It is the number and range of links between nodes that will determine how resilient an ecosystem is to change and how "collapse" events might lead to future restructuring. In parts of Africa the increasingly negative effects of climate change have already reduced fruiting of trees that many megafauna depend on.[36] Fruit famine, due to rising temperatures and increasing drought, might reasonably trigger population collapse of forest elephants, great apes, monkeys, and myriad bird species. But the functional roles played by these species are complex, diverse, and hard to quantify. Will faunal losses such as these cascade into other species who depend indirectly on the ecological engineering of those vulnerable few?

Vertebrate extinctions are certainly emotive, and evidence of precipitous decline of individual species abounds. But invoking catastrophic collapse on the basis of a few can be misleading. The *2020 Living Planet Report* from the World Wildlife Fund found that, since 1970, more than half of all vertebrates have been lost, a staggeringly rapid and worryingly significant scale of decline. Yet this "average" is skewed by very large declines in only 3 percent of the vertebrate populations considered. When those are excluded, the global trend for the remaining 97 percent is one of increase, rather than decrease.[37] To avoid the bias introduced into these kind of average statistics, a more meaningful and accurate representation can be gleaned from viewing the data in terms of clusters. By measuring the trend of change for clusters of species, this alternative approach shows that although 1 percent of populations have experienced extreme decline and about half as many experienced extreme increase, the very great majority showed no trend at all.

This isn't to say that there is no reason for concern. On the contrary, those species that have declined so significantly are in the gravest of danger, and perhaps here lies the true value of this more focused analysis. When faced with news of the "sixth mass extinction," it is tempting to lose hope, to assume that all life is imperiled and that the enormity of the crisis is so vast that no meaningful action is possible. But that is not the case. The vast majority of life on Earth appears, for the moment, to be doing fine, but there are a number of ecological systems—groups of interacting species or functional types—that are increasingly vulnerable to collapse. The need for action is real, and targeted conservation is urgent. The evidence for particular thresholds beyond which catastrophic damage is inevitable is still lacking, and in the meantime such a framing tends to underrepresent the very real threat exerted by small but incremental environmental change.

Above all, the historical and fossil record reassures us that, to some extent, many species can survive changes in climate. But to do so they need the appropriate functional habitat and the connectivity of habitat to facilitate migration. A bold conservation plan is necessary, one that brings about large-scale transformation of the global food system in ways that ease the unbearable agricultural pressure on our natural ecosystems.[38] In doing so we simultaneously gain the additional benefit of reducing the emergence of human-affecting animal-borne infectious diseases that are known to be linked to the conversion of natural to agricultural habitats.[39]

By reshaping the natural hierarchy in ways that have placed us deliberately at the apex, we have assumed superiority in many forms, yet we choose to recognize these forms selectively. What we are learning, through Newtonian scientific approaches, is something that indigenous cultures have perhaps known for centuries—that to survive as individuals and as a species, we must be stewards of our total environment.[40] Stewardship means taking action to mitigate threats across the ecological spectrum, not simply those that represent socioeconomic challenges for our one species. Full-cost accounting is necessary to balance the ecological-financial books, to appropriately value the worth of future benefits, and to quantify their relationship to the integrated cost of short-term biodiversity losses. And although such a challenge may appear intractable, our history tells us that we are certainly capable of healing even large "self-inflicted wounds."[41] Observation, monitoring, and concerted international efforts have allowed us to largely restore the ozone hole, to reverse the near-extinction of certain whale species, to clean our air sufficiently that sulfuric-acid rain is now far less prevalent than it was only a few decades ago. The tragedy of the commons is not an inevitability, but it is a warning that needs to be heard—the freedom of our "superiority," if that is what it is, must be matched with a commensurate level of societal responsibility. The invisible hand of evolution may instruct and guide the orchestra of life, but the musicians must know the score, await their turn, and temper their exuberance.

Life on Earth emerged when the planet was geophysically ready for it.[42] Photosynthesis harnessed energy for complex life, and through cascading gradient-driven flows, this energy provided the drive that fueled the adaptive cycle of natural selection. Optimization of life through cycles of speciation and extinction arose because energy flowed through the connected web of global ecosystems, just as fungal threads connecting the roots of forest trees use electrical pulses and scale-free subterranean networks to communicate. Nodes within these networks are connective hubs that ensure species interaction, but without functional redundancy, these networks are vulnerable. Modularity, in which unconnected sub-systems coexist, brings resilience and slows the cascading collapse of interdependent ecosystems. As we, the human species, begin to make deliberate decisions for future ecological engineering, such concepts must be foremost in our collective minds.

Because right now, more than 70 percent of us live in places where our consumption exceeds that which can be supported sustainably by the environment. We increasingly support ourselves and the infrastructure that underpins our societies on energy derived from the sun three hundred million years ago—energy in the form of coal, oil, and gas. Perhaps the ultimate time-delayed feedback is the one that we are now rapidly racing toward—the release of fossil energy into a newer world unable to absorb or dissipate it quickly enough. The consequences will certainly follow the pattern we have seen repeatedly so far, one of an initially slow start but then an accelerating rate of change that persists until a new equilibrium is established. We are now taking our first steps on the nursery slopes of a critical transition, a steepening trajectory that will redefine for us and all other species the very thing that keeps us all alive. We stand at the threshold of a new global *Climate*.

CLIMATE 3

Already he knew how to recognize Atman within the depth of his being, indestructible, at one with the universe.

—HERMANN HESSE, *SIDDHARTHA*

W hen the *Sixth Assessment Report* of the Intergovernmental Panel on Climate Change (IPCC) was published in August 2021, the international media storm was intense. The result of more than three years of exhaustive investigation and four successive independent review periods involving thousands of experts all over the world, the million-word document went public just three months before COP26 in Glasgow, the United Nations climate change conference more formally known as the 26th Conference of the Parties. The secretary-general of the United Nations, António Guterres, described the report as "code red for humanity." But what caused the furor wasn't the underlying message of the report— that the global climate is changing and that human activity is the reason—it was the words with which that message was conveyed that shocked the world. When the *First Assessment Report* was released in 1990, the evidence available was insufficient to establish, with any certainty, our role in changing the climate. Now, three decades and five assessment reports later, the findings were stark. Evidence for human influence on climate was described as "unequivocal," and the scale of impact "unprecedented." To the 234 authors of the report used to working on the front lines of climate science, these conclusions came as little surprise, but the evidence, compiled by experts spanning a vast range of disciplines painstakingly distilled into sound bites for the executive summary, was nonetheless sobering.

How did we get into such a state, and how could things have gotten so bad so quickly? In the 1990 report, the authors set the scene by remarking in the first line of their introduction that there was "concern that human activities may be inadvertently changing the climate of the globe." As if we hadn't known that this might happen. As if this was a terrible accident that couldn't have been foreseen. But history is not on the side of those who might claim ignorance. As far back as 1856, scientist Eunice Foote was conducting experiments showing that air with additional carbon dioxide absorbed more heat than air with less carbon dioxide and, most critically, that it then took longer to cool.[1] Later, in 1938, Guy Stewart Callendar published a paper in the *Quarterly Journal of the Royal Meteorological Society* titled "The Artificial Production of Carbon Dioxide and Its Influence on Temperature," in which he calculated human-caused carbon dioxide emissions and their theoretical effect on global temperature.[2] Comparing those estimates to observations from two hundred weather stations, Callendar proved Foote had been correct and, in doing so, confirmed nearly a century ago a simple truth that we are still reluctant to acknowledge. In a 1975 paper in the journal *Science*, Ruth Reck showed that the carbon dioxide and aerosol emissions produced by vehicles could produce atmospheric warming at Earth's poles—research that she conducted while working in the physics department of General Motors Research Laboratories.[3] Three years later, and forty years after Callendar's paper, glaciologist John Mercer published research in the prestigious journal *Nature* in which he referred to the observed carbon dioxide greenhouse effect as a "threat of disaster."[4] The climate crisis we currently find ourselves in was therefore most certainly predicted and avoidable. We are here simply because we gambled with a frighteningly powerful feedback we thought we could perhaps ignore, one that it now turns out we cannot.

The Makings of a Carbon Time Bomb

The story of our current and future climate is therefore one of decisions made in our relatively recent history, but the origin of this story literally lies beneath our feet. The carbon dioxide climate feedback is perhaps one of the most simple and yet devastatingly disruptive feedbacks we are able to witness. It is a feedback whose time delay spans hundreds of millions of years, stretching back to times in Earth's geological past when vast tropical forests covered continents whose tectonic wanderings had positioned large swaths of them at tropical latitudes, close to the equator, in climate regimes that favored the luxuriant growth of dense and uninterrupted swampy forests. This period, the Carboniferous, takes its name from that element whose capture and fossilization was the defining characteristic of the time. Trapping energy from the sun as these plants flourished, they locked away for hundreds of millions of years the warming potential of that atmospheric trace gas that had once been so much more abundant. Respiring deeply, patiently, long breath in, then slowly out, the prolific growth of lignin-dominated plants drew in carbon dioxide from the air. And because lignin resisted decomposition, the once-gaseous carbon eventually became entombed in solid form, interred in thick organic layers whose weight compressed and consolidated but did not destroy. With so much carbon being removed from the atmosphere, the relative proportion of oxygen in the atmosphere began to rise,[5] perhaps fueling the evolution of gigantism in insects and amphibians that were previously limited in size by their ability to respire.[6] The balance of the breath of life reweighted, adjusted, realigned for another experiment. Gradually sequestering carbon dioxide from the air, the Carboniferous forests were so organically dense that their now-fossilized remains contain more elemental carbon than anything else. And it is this carbon, in a hydrogen and sulfur compound, that when oxidized by combustion, has during the last 250 years liberated from those ancient deposits such an abundance of fossil energy that it has fueled a monumental transformation in our global population, changing our societies, behaviors, technologies, and agricultural productivity. But that energy, the breath of ancient life, has at the same time been polluting our atmosphere and acidifying our oceans, threatening to trigger a new wave of abrupt environmental change the nature of which we are still only just beginning to see.

So how did it happen that plant growth in the tropical swamps of Carboniferous Earth, the silent lungs of Gaia whose whisper once encouraged a new era of life, became the basis of the three-hundred-million-year-delayed climate feedback we are now living with? How is it that complex life was able to evolve and flourish within the bounds of a natural climate that oscillated cyclically in a regular and predictable manner for hundreds of millions of years, only to then set in motion in the last 250 years a cascade of feedbacks that have already changed the global climate to such an extent that we have effectively prevented the next ice age?[7]

In short, industry. Specifically, the advent of mechanical devices that required a continual and dependable source of power. When the IPCC released its latest report, it confirmed that global average temperatures had reached 1.1 degrees Celsius above "preindustrial" levels, a moniker that is commonly used to identify a time before mass mechanization and coal-driven steam power elevated the levels of carbon dioxide in our atmosphere. But although this term is widely adopted, it has only recently been specifically defined. The problem is that coal-fired industry didn't just appear in an instant; it emerged first in eighteenth century Britain before spreading across Europe and the rest of the developed world in subsequent decades. And at first, the scale of coal burning was small, so the impact on the climate was undetectable. Not only that, but there also weren't many meteorological observing stations recording regular climatological measurements at that time, making it even harder to establish how things have changed since. The early industrial period of the late eighteenth century marked the beginnings of coal-powered industry, and so the true "preindustrial" period really refers to times before this.[8] But it is only from the mid-nineteenth century that continental and marine temperature records reveal a robust uptick in their readings.[9] This, then, is when the Earth first felt the impacts of that Carboniferous climate feedback.

The origin story of the Industrial Revolution is complex and uncertain. Like many emergent phenomena, there were a multitude of small developments and innovations that at first simply fizzled and disappeared without ever amounting to much. *Long breath in, slowly out.* A regular rhythm of experimentation and progress, human aspirations kept in check by the regulating limiter of technological possibility. But when, in the late eighteenth century, Scotsman James Watt improved the earliest steam engines, he made such effective changes that this new engine revolutionized the efficiency with which, for example, textile producing machines could operate. The textile industry was at this time a major source of employment in Britain, and the goods produced contributed substantially to the national economy. By coupling mechanized "spinning jennies" to spool yarn more quickly and efficiently than previous manual-powered versions with mechanized looms to weave that yarn into cloth, coal-fired steam power quickly supplanted traditional methods of textile production. Using a single machine to replace the work of entire teams of spinners and handloom weavers, the drive for optimization, efficiency, and, above all, profit manifested as a social feedback that changed the working patterns and economic dependencies of thousands of workers in the counties of northern England. From a low-productivity, spatially distributed cottage industry carried out across the agricultural regions, textiles became a high-productivity, spatially concentrated industry accommodated in the new factories springing up in cities like Manchester, Nottingham, and Derby. Together with advancements in the design of machinery, and the availability and transport of raw materials and labor, innovations such as Watt's paved the way for the Industrial Revolution of the eighteenth and nineteenth centuries. But the advent of mass mechanization across Britain was only possible because it made such effective use of coal, and from the coalfields of northern England, this was a fuel that was plentiful. Gaia's lungs, now blackening. With abundant fossil fuel at the disposal of these early industrialists and an exponentially increasing global demand for power, a regime shift took place through the eighteenth and nineteenth centuries that exerted a feedback so strong it irrevocably changed the trajectory of global population growth and the economic productivity of the rapidly industrializing nations. But it did something else too. This rush for coal-powered mechanization increasingly separated us from the natural environment that, for seven million years, had nurtured and enabled us.

Measuring the Breath of Life

Today we measure carbon dioxide in the atmosphere relative to its concentration during these early industrial times. In the ten thousand years following the end of the last ice age, the partial pressure of carbon dioxide in the atmosphere increased by only one hundred parts per million—a difference that nonetheless helped warm the ice age world by an average of 7 degrees Celsius and allowed the current "interglacial" climate to arise.[10] But in the 250 years since the first experiments with coal-powered steam engines, our climate has seen another 140 parts per million of carbon dioxide added to the skies from our accelerating fossil fuel combustion. Despite twenty-five years of global climate policy negotiations, carbon dioxide has continued to increase, with implied future warming that will radically change every aspect of the world we live in. How do we map this situation onto a wider framework that can show us, objectively, how these changes compare to changes we know have occurred before? How might we hold a stethoscope to the lungs of our Earth and listen, earnestly, to our patient? *Long breath in, slowly out.*

Today we can measure carbon dioxide in the atmosphere accurately and chart the seasonal rise and fall of its concentration as the deciduous forests of the Northern Hemisphere lose their autumn leaves and fall dormant, only to grow new ones each spring and hungrily take up that gas that gives them life. This oscillation from winter to summer amounts to a variation of about six parts per million. If nothing else changed, atmospheric carbon dioxide levels would stay more or less constant, gently fluctuating with the seasons as the floral lungs of the Earth inhaled and exhaled, in then out, slowly and rhythmically like a meditation. In Sanskrit the word *At-*

man is given to the self, the real part or essence of a being, the observing pure consciousness. But it also translates as "breath." And so it is that the fundamental essence of Gaia, the eternal, radiant, and divine nature of the Earth is intimately associated with the breath of life that we transcribe from the air with a precision that captures changes as tiny as one thousandth of a part per million, one gram in a thousand tons. Against this divine blueprint of the fundamental self our measurements show irrefutably that in the last fifty years the annual average concentration of carbon dioxide in our atmosphere has increased by ninety-two parts per million.

Charting the recent changes in our atmosphere is easy because we can directly measure the concentrations of the various gases in the air. And we can compare these changes to measurements of temperature that we can also measure directly, both in the air and in the ocean. But the techniques on which these measurements are based are relatively new. Carbon dioxide, for example, was only first directly measured in a regular and standardized way in 1957, during the International Geophysical Year. Charles David Keeling, a postdoctoral researcher at Scripps Institute of Oceanography at La Jolla, California, took gas readings from the summit of the extinct Hawaiian volcano Mauna Loa four times a day, every day, for eighteen months. In doing so, he showed conclusively not just that atmospheric carbon dioxide concentrations were higher than the measurements made by Guy Callendar twenty years before, but also that the concentration was measurably increasing, even within the relatively short eighteen-month period of new measurements. Devices to determine temperature originated considerably earlier, perhaps in the late sixteenth century with Galileo. But it was only in the early 1700s that the German physicist Daniel Gabriel Fahrenheit invented the mercury-in-glass thermometer and the standardized temperature scale that we still use today.

To deduce anything of atmospheric or climatological changes older than the last few hundred years, therefore, we have to rely on either the extraction of tiny bubbles of ancient air captured and preserved in ice, or on climate proxies—environmental indicators that can be used to calculate past climatic conditions based on relationships shown to accurately relate changes in a given proxy with known changes in the quantity it reflects. Today, there are myriad techniques employed by scientists all over the world that can extract indications of age from a vast diversity of materials. Radiocarbon dating is well known, relying on the rate of radioactive decay of specific isotopes of carbon to determine when a formerly living organism died. *Long breath in.* More esoteric are approaches like optically stimulated luminescence dating, electron spin resonance dating, cosmogenic nuclide surface exposure dating, or fission track dating, which can literally retrieve ages from solid rock. But until the nineteenth century, no one had needed such techniques, because for the most part, the idea that the climate might have changed through history was something associated more with myth and divine intervention than it was with known physical principles, such as the Newtonian mechanical view of the solar system. In the early decades of the nineteenth century, however, that began to change.

In 1830, the Scottish geologist Charles Lyell published the first volume of his three-volume *Principles of Geology*, which among other things put forward the idea that the "present is the key to the past," that is, that the Earth evolved in the past according to the same physical laws that apply today.[11] This so-called uniformitarianism starkly contrasted with the idea of "catastrophism" that had been previously popular, in which Earth's passage through time was marked by catastrophic events that wrought change through cataclysm and destruction. The more gradualistic view proposed by Lyell, however, allowed far more insight into the past than catastrophism offered. By translating observations from present environments onto past landscapes, nineteenth-century geologists gained considerable new abilities in the interpretation of the past. Using precisely this approach, one of the first to comment in the scientific literature that Earth's climate may have differed from its present state in the relatively recent geological past was Swiss naturalist Louis Agassiz. Inspired by theories emerging in the 1830s that described how the glaciers seen at that time in the European Alps could shape and modify the landscape, Agassiz began recognizing the telltale signs of glacial land systems elsewhere in Europe, most notably in the mountains of the British Isles. Agassiz and others quickly established a theory that vast ice sheets had once covered much of northern Europe and North America.[12] Since the climatic conditions required for glaciation were considered to be significantly different than those prevailing during the nineteenth century, the implication of these conclusions was that the Earth was previously much colder, that great ice ages must have once enveloped the globe.

At around this time, another Scot, James Croll, became interested in Lyell's ideas about the role of Earth's orbit in bringing about the hypothesized global ice ages. Croll had little in the way of formal education but while working as a caretaker at what is now the University of Strathclyde, Croll used his access to the university library to teach himself physics and astronomy. Based on what he read, Croll proposed that Earth's changing orbit around the sun might affect how the ocean currents of the North Atlantic transferred heat to the Arctic. At times when this heat flux was reduced, sea ice would expand due to the cooler sea surface, and in doing so, the bright snow and ice would reflect more sunlight than the dark ocean previously had, further cooling the climate. In short, Croll put forward the idea that the "gradualist" changes in Earth's orbit could trigger climatic changes that led to a positive feedback loop of cooling and the origins of a global ice age.[13]

This remarkable insight set the tenor of the next 150 years of climate research, and in many ways, the findings of the *Sixth Assessment Report* are built on Croll's foundational deductions from the mid-nineteenth century. But for many decades after its initial publication in 1864, the orbital control of Earth's climate remained speculative, and by the time of Croll's death in 1890, his theory had been largely dismissed. The whispering breath of life was too quiet, and the world was not yet ready. Yet he was right, and it was a Serbian mathematician, Milutin Milanković, who proved it. Bringing mathematical rigor to Croll's intuitive discovery, Milanković calculated the orbital influence on Earth's climate precisely, isolating the deviations in solar energy received by the Earth as its orbit around the sun became more elliptical ("eccentricity"), as Earth's tilt changed ("obliquity"), and as the angle of Earth's rotation about its axis shifted ("precession"). Although these components of orbital and axial variability had been known about in some cases since Greek times, it was Milanković who derived the combined effect of these 100,000-, 41,000-, and 23,000-year oscillations in terms of what that meant for the amount of incoming radiation, or insolation, that affected the Earth's surface. Armed now with a concrete assessment of the time-varying energy budget of the Earth, it was then possible to predict with some certainty the rise and fall of the ice ages. And yet, somewhat remarkably, it was only fifty years ago that the geological evidence for these ice age cycles was finally discovered. Using seabed drilling technology to recover sediment cores from the deep ocean, the National Science Foundation project CLIMAP (Climate: Long Range Investigation, Mapping, and Production) showed that during the last half-million years the Earth's climate had indeed changed for the exact reasons predicted by Croll and subsequently calculated by Milanković. This discovery, referred to as the "pacemaker of the ice ages,"[14] marked the beginnings of the modern era of palaeoclimatology and established the framework we still rely on for the interpretation of the changes in climate we are beginning to see today.

Prediction Problems

Several key elements of the processes governing long-term changes in climate had, by the end of the twentieth century, fallen neatly into place.

Orbital variability controlled the amount of sunlight reaching the Earth, how it was distributed between the hemispheres, and how the intensity of the incoming energy was spread out through the year. The different periodicities and energy content of these fluctuations had been accurately calculated, and the resulting predictions of ice ages had been borne out by evidence from seabed sediments. Yet there was a problem. When researchers reconstructed ice age climate changes of the last half-million years, they saw clear evidence of cycles spanning one hundred thousand years each.[15] This frequency corresponded neatly to the timing of changes in the elliptical shape of Earth's orbit, or eccentricity. Over one hundred thousand years or so, Earth's orbit around the sun flexes slightly due to the gravitational pull of other planets in our solar system. As a result, our orbit becomes at one extreme a little more circular and at the other extreme a little more oval. And it is during the periods of high eccentricity (more oval orbit) that the greatest seasonal differences in incoming energy occur. But because the seasons in each hemisphere oppose one another, the variations in the globally averaged total amount of energy reaching the Earth are disappointingly small. Over this one-hundred-thousand-year cycle, the variability of total annual energy across the Earth's surface amounts to only about 0.35 watts per square meter, or about 0.1 percent of the annual total. How could such tiny changes in insolation over one hundred thousand years result in global ice ages spanning the same length of time? This "hundred-thousand-year problem" has dogged the field of paleoclimate for decades, with numerous attempts having been made to explain it. The problem is made more complex by the fact that, up until around eight hundred thousand years ago, the ice age cycles seen in marine sediment sequences tended to show much shorter periods—typically the forty-one-thousand-year frequency more closely related to changes in Earth's tilt (obliquity). If Croll was right about Earth's orbit controlling the pacing of the ice ages, then how could the frequency of this pacing have changed from forty-one thousand years to one hundred thousand years, and why would it do so, given that the energy variations at one-hundred-thousand-year frequencies were so minuscule?

This transition in ice age frequency took place gradually over at least several hundred thousand years and has come to be referred to by climate scientists as the "Mid-Pleistocene Transition." What was it that happened at this time? Although geological records provide valuable indications of the cyclical oscillations of the climate, they unfortunately do not readily offer clues as to the mechanisms that caused them. For that, we need computer models that encode mathematical relationships into algorithms and calculate incremental changes over and over again until each equation is solved.

When the ancient Greeks built the Antikythera mechanism in the second century BC, they were building a tool that represented the observable sky with sufficient accuracy that they could make predictions regarding the timing of lunar phases, solstices, or planetary movements. The bronze gearwheels that controlled the rate at which each celestial element rose or fell in the simulated night sky were connected to one another, such that the entire population of astronomical bodies moved in concert, tracing choreographed paths on a heavenly stage. The precision of this first computation machine was remarkable, and its users could, with reassuring fidelity, quickly explore planetary configurations both past and future with relative ease. One of the most useful assets of the astronomical system that the Greeks explored was that it was highly regular and dependable. The observer could see how the different periods of axial or orbital rotation of each separate body might yield a very specific and perhaps even unique configuration of the whole group, but the pieces of the puzzle were rigidly connected and unintended interactions were not possible. As a model, therefore, the Antikythera mechanism was a wonderfully illustrative device that gave results that were always predictable and unsurprising. And this is where climatology differs so clearly from astronomy. The weather that we each experience every day, in every corner of the globe, is a very rapidly evolving system with temperature, rainfall, and winds able to change abruptly from one minute to the next. And each of these changes can trigger a different result elsewhere in the world—warming in one place brings rainfall to another or drives changes in atmospheric pressure that alter the path and speed of winds. Climate is simply the average of weather over several decades. And compared to planetary changes, the difference is that the links between components of the weather or climate system are neither rigidly fixed nor uniform in strength and direction through time. Rather, the connections depend on the state of the system, and this state arises *because* of the connections.

In the middle part of the twentieth century when climate scientists first started using computers to speed up the solving of routine atmospheric equations, they discovered firsthand the importance of these "state-dependent" connections and, over the course of just a few years, revolutionized science and gave birth to a whole new field—*chaos*. It was 1961 and Ed Lorenz, then an assistant professor at Massachusetts Institute of Technology, was beginning to recognize that the traditional "linear" models of climatic behavior were wrong.[16] Linear models are those in which a change in one quantity leads to a proportionate and predictable change in another. In a linear scheme, if you start your calculation with a slightly different value from a previous simulation, you just end up with a result that is also slightly different. But what Lorenz discovered as he computed how temperatures within a column of air would evolve through time was that the result he produced could be completely different, not just slightly different, if he started with even just a tiny change in the starting value of one of the terms in the equations. What he had discovered was the *nonlinear* response of the atmosphere, a property that led Lorenz to show for the first time that the climate can behave chaotically, abruptly switching between markedly different states even without being "pushed" by some form of external force. We already knew that the climate was a highly complex system composed of many elements that all interact over different time periods. Some interactions could be very rapid, such as the freezing of water in the air when the temperature drops to a specific and predictable temperature. Others might be much slower, such as the interactions between air masses over the oceans and the circulation of the deep water beneath. But what chaos theory now showed us was that, because the climate system is so intimately coupled, its large-scale behavior becomes increasingly unpredictable the further ahead we try and look. A tiny and seemingly insignificant change to the initial state becomes rapidly amplified through nonlinear feedbacks and results in an outcome that is no longer just a consequence of the climatic conditions at the start, but over time becomes ever more a reflection of what has happened *since* the start. The rapidly repeating but ever-changing feedbacks between climate components, coupled with the possibility of chaotic behavior, means that new climate states might "emerge" with little warning or under conditions not previously anticipated.

And so if we return to our "hundred-thousand-year problem," that transition from forty-one-thousand-year to one-hundred-thousand-year climate cycles, we now have a new perspective. If a chaotic system can "jump" between states even without being pushed from the outside, could this explain how Earth's climate flipped from one that responded to changes in the tilt of the planet's axis to one controlled more by the shape of its orbit around the sun? Thought experiments and mathematical curiosities are one thing, but how do we unravel the actual physical processes, the feedbacks that connect them, and the cascades of nonlinear behaviors that must have been necessary to bring about such a large-scale shift in planetary functioning?

One approach to tackling such a complex problem is to employ a complex model. Climate reconstructions of the past, just like the future projections of the kind presented in the IPCC reports, rely for the most part on general circulation models (GCMs). A very simple GCM might simulate only the atmosphere, solving long series of equations just as Lorenz had done, but simultaneously attempting to use the laws of physics to calculate the evolution of all of the major components of the climate—temperature, humidity, air pressure, rainfall, wind speed, and so on. GCMs do this for the whole globe by dividing it into a grid of "cells" that together span 180 degrees latitude and 360 degrees longitude. At each of the locations on this grid, all the equations of state are solved for a given point in time, and these solutions are then used as the basis for the exact same equations at the next point in time. Gradually the system evolves, because the solutions to the equations at each successive point in time depend on the state of the system at the previous point in time, each component influencing the others to a greater or lesser degree according to the strength and direction of the feedbacks involved. This hideously complex atmospheric system gets yet more complicated, however. Each of the grid points in the horizontal plane is also replicated in the vertical axis, as well, because the atmosphere of course isn't just a single layer of air. Heating the land surface causes the air to warm and rise up, through convection, taking with it moisture that eventually cools and forms clouds. So the "computational box" of this atmospheric model has around forty vertical levels as well as all those horizontal grid points. Multiplying these dimensions, we can see that at every time step taken by the atmosphere model, it needs to solve its innumerable suite of equations for a multitude of different quantities about two-and-a-half million times ($180 \times 360 \times 40$) and then store the answers to those calculations to use as inputs to the equations at the next time step. But although it is the atmosphere that we think of most often when referring to "climate," its behavior and evolution are inextricably tied to the other Earth system components that it interacts with. The global oceans move heat and salt around the planet, and just like the atmosphere, the pattern of their circulation also has to be simulated in three dimensions. Warm water from the tropics moves poleward in both hemispheres and transports heat to higher latitudes. As the water cools, it becomes denser and sinks through the water column. Combined with increasing salinity as polar sea ice locks up freshwater and rejects brine, the water at high latitudes forms denser currents that then flow back, at depth, to warmer regions, where the cycle begins again. Today, GCMs have swelled in sophistication so much that the more complex ones are now commonly referred to as Earth system models (ESMs). These models not only simulate the atmosphere and ocean, but also the growth, movement, and decay of sea ice; changes in the water content of lakes and rivers on land; vegetation changes and land-use patterns; biological productivity in the ocean; the growth and flow of ice sheets; and even the emissions from human industry and agriculture. Most importantly, ESMs track the movement of carbon and other nutrients through the entire global system, allowing for chemical, biological, and physical feedbacks to develop between each and any of the interconnected components—an ethereal Gaia for the digital age.

Mechanistic climate models such as these, which have sought to understand the combination of physical processes that led to the "switch" in climate behavior during the Mid-Pleistocene Transition, are numerous, but unfortunately they often differ in how they see the relative importance of each of the components of the Earth system. In some models, the structure or amplification of ice age climate cycles mostly depends on the amount of carbon dioxide or other greenhouse gases in the atmosphere,[17] whereas in other simulations carbon dioxide is considered less important than slowly oscillating variations in solar energy.[18] Most studies agree, however, that additional climate feedbacks must have amplified the changes taking place, and these feedbacks could have come from myriad sources. Dust in the atmosphere, carried aloft from desiccated and frozen landscapes would affect how much sunlight reaches Earth's surface,[19] whereas the slow subsidence of the Earth's crust due to the weight of growing ice sheets would lead to a time-delayed feedback in how climatic changes first accumulated then melted snow and ice.[20] From previous chapters, we might recall that the erosion of rocks and sediments from the land beneath the ice sheets could impact oceanic conditions, changing their biological productivity and altering the absorption of atmospheric gases. Not only that, but by eroding sediments, these ice sheets also affect how they themselves behave, building up or stripping away the lubricating layer of sediments that helped them flow more expansively across the landscape.[21] Beneath the changing atmosphere, the ocean slowly but progressively cooled and warmed while sea ice at its surface grew and decayed much more rapidly and the water column switched repeatedly between states of either stratification that isolated the deep ocean from the atmosphere or convective overturning that mixed the layers of water and the heat they contained.[22] Maybe all of these processes were important, but often the conclusions reached from any given study have tended to reflect the processes that were included, or excluded, from the model and the way that the model was set up. When each model tells a different story, then, whose should we believe?

Perhaps we need to believe them all. But not in the way that they might have originally been presented. For the breath of life that regulates our planet and gives life to the phenomena upon it is one that brings cyclic dependence despite an ever-changing environment, uniformity despite catastrophe, and growth and adaptation despite the near constancy of our astronomical situation. So to make sense of conflicting claims, we need to reframe those answers around a reimagined question. Instead of asking how the climate changed at any one time in the past, we could instead ask why the climate changes of the past were, at least within the last half-million years, so remarkably consistent in their pattern. What clues might that pattern yield to us so that we can better understand the feedbacks, and make better predictions for the future?

Let's start by defining the outline of what we know, the history of our climate through the most recent periods of Earth's past. To do this, we turn to environmental indicators—proxies—laid down in ocean sediments during the last few million years, the alternations of which first proved fifty years ago that Milanković, Croll, Agassiz, and Lyell were right in their astronomical theory of the ice ages. From sediment cores recovered from the depths of the major ocean basins of the world, a remarkably clear story can be told, a reconstruction of changes in Earth's temperature spanning the last sixty-five million years.[23] Yet for much of that time, the global landscape was quite different from today, with continents arranged differently and ocean circulatory patterns correspondingly altered. Focusing on the last five million years, however, from the beginning of the Pliocene (5 million to 2.8 million years ago) through the Pleistocene (2.8 million to 11,000 years ago) and into the present "interglacial" climate of today, a picture of Earth's natural climate variability can be drawn that is less affected by the tectonic wanderings of deeper geological time. Bringing together sediment records from fifty-seven globally distributed seafloor archives, geologists Lorraine Lisiecki and Maureen Raymo published in 2005 what has since become a monumental keystone for paleoclimate reconstruction.[24] By using the same kind of multi-record approach as for the sixty-five-million-year temperature reconstruction, Lisiecki and Raymo aligned each of their individual datasets to the same timescale and "stacked" them to better resolve the global climate signal and reduce the influence of local changes. From this, they showed that the chemical composition of seawater (reflecting water temperature as well as global ice volume) oscillated in cycles of around forty-one thousand years before about one-and-a-half million years ago and then in cycles of around one hundred thousand years from approximately a half-million years ago until today. In the intervening period, the climate appeared to undergo some kind of reorganization in which the dominant period of fluctuation gradually changed from the shorter (obliquity) frequency to the longer (eccentricity) frequency. But this gradual change was not one of a slow increase in the length of each ice age cycle. The three so-called Milanković cycles were always present; the change was simply one of dominance—eccentricity "replaced" obliquity when the influence of the longer period became sufficiently great to overprint the shorter period.

To achieve their remarkable and almost continuous history, Lisiecki and Raymo had to make a few assumptions and to sacrifice a degree of accuracy in order to tell the most complete story. So although their reconstruction depicted with impressive fidelity the tussle of Earth's climate system and the general pattern of change over millions of years, the approach they had adopted to remove the "noise" and reveal the signal had left them with a portrait of the climate in which chaotic changes of the kind predicted by Ed Lorenz were notably absent. Thankfully there are other sources of information that can reveal precisely the kinds of atmospheric changes that ocean sediments simply cannot. Since the late nineteenth century, glaciologists had tried drilling into glacier ice to investigate the formation and structure of those features. But it wasn't until the late 1950s that drilling technology was being specifically designed not just to drill into the ice, but also to recover a solid core of the material that could be analyzed in a laboratory. Recognizing that the great ice sheets of Greenland and Antarctica held the thickest ice and thus potentially the longest record of snow accumulation, the International Geophysical Year of 1957 to 1958 saw drilling projects simultaneously trying to recover ice cores from both ends of the planet. As technology and experience grew, the length and quality of the cores retrieved increased rapidly. Today we have dozens of cores from the two main ice sheets, which in some cases extend more than three kilometers in length and span the last eight hundred thousand years and offer far more detail than marine sediments. And there's something else about ice cores that makes them the de facto choice for climate reconstructions of the last nearly one million years. Whereas marine sediment records relied on changes in seawater chemistry trapped in the preserved remains of ocean-dwelling organisms to draw inferences of past climatic change, with ice cores the atmospheric signal was captured directly. As snow fell on the ice sheet surface, the layers it formed recorded the chemistry of the air in which it was formed. The balance of different isotopes of oxygen, carbon, and nitrogen varies through time as temperatures change and varies geographically depending on where the air mass originates. When these layers of ice are melted, the isotopic changes can be measured with great precision and converted into quantities like air temperature or the rate at which snow accumulated, which tells us how wet the air was. In the upper layers of the ice, before the pressure of ice above becomes too great, bubbles of air are trapped in the ice as it gets progressively buried. These bubbles are time capsules containing tiny samples of the atmosphere just as it was, at, or shortly after, the time when the snow in that ice layer fell from the sky.

During the last two or three decades, these ice core "time machines" have yielded ever-clearer images of our past climate and how it changed from one year to the next, one ice age cycle to the next, back to the time when our Earth was beginning to breathe more slowly, inhaling and exhaling to the mantra of the one-hundred-thousand-year cycle. And from these ice core records, we have seen for the first time exactly the kind of abrupt and seemingly chaotic change that Lorenz had imagined more than a half-century ago. Layers in the ice become less easy to distinguish as we go deeper into the core, further back in time, but as our analytical capabilities grow ever more sophisticated, we can isolate individual seasons within a year in the younger parts of a core or the individual years and decades as we head back to the end of the last ice age, eleven-and-a-half thousand years ago. At this time ice from Greenland tells us that during just a few decades the climate there warmed abruptly by up to 10 degrees Celsius, terminating the ice age and marking the onset of our current mild interglacial.[25] But just as Lorenz had predicted, the chaotic climatic system didn't just show a simple step change in temperature during this termination, it showed many large-scale fluctuations repeatedly jumping between cold and warm states, until eventually a new stable state was found. Further back into the glacial period we see other abrupt jumps, some of which are only seen in the Greenland ice cores and some that appear synchronously with changes observed in Antarctic ice. Collectively we can assemble these traces of our ice age climate and begin to separate local from global events, transient blips from long-lasting deviations, chaotic behaviors from astronomical forcing. Among this cacophony of information there are clues to the feedbacks at work and the connections that waxed and waned through time, the mechanisms that shaped the evolving climate and collectively wrested periodic regularity from the maelstrom of nonlinear behaviors.

When Less Is More

From Gaia, according to legend, was born Uranus, the sky. The goddess of Earth wanted to be enveloped on all sides by the heavens to be whole and complete. She bore mountains and oceans, and in the unity of these realms, Gaia was the ancestral mother of all life. The breath of Gaia is not, therefore, a singular action, but a process, a means of living, a recirculating flow that invigorates and enlivens the entirety of our planetary system. In perceiving her overwhelming vastness, we are daunted by her complexity, the chaotic dance of the weather incessantly executed across an ever-changing terrestrial stage. How then to entice our newly constructed digital representations of this goddess to reveal to us, the fallible mortals and clumsy empiricists, her divinity, her beatific orchestration of *omnis terra*, the whole Earth?

Numerical climate and earth system models have provided increasingly clear insights into climatic change during periods of the past as well as under scenarios for the future. But their astounding complexity is also their Achilles' heel. For the computational resources required to run these models routinely demand access to a supercomputer, and as the models and simulations get more and more complex, the computations demand the very fastest machines, exascale supercomputers capable of carrying out a billion billion calculations per second. Yet even with such powerful tools, performing simulations spanning hundreds of thousands of years remains intractable. How then might we try and understand the significance and consequence of global climatic feedbacks over very long time periods?

The first step is to acknowledge that although mechanistic models excel at reproducing particular processes, when too many processes are combined, the resulting feedbacks between all those processes tend to make the model unstable and uncontrollable. "Beyond a certain point," says philosopher Alan Watts, in considering the human mind, "the mechanism will be 'frustrated' by its own complexity."[26] So too with our models. To regain control we have to smooth out any abrupt changes, but if the climate system really is chaotic, those abrupt changes are exactly the kind of behaviors we need to study. To understand the behavior of the whole system, therefore, we might instead adopt an approach that seeks to first mimic in mathematical form the pattern of climate change seen in the ice core and marine sediment records and then to work outward from that mathematical representation to learn something about the ineffable nature of the goddess controlling the system, without questioning too closely the processes she employs. With such models our ambition cannot be to reproduce "reality," because our equations are too simple. Instead we hope only to draw back the curtain a little, to illuminate anew the patterns and style of the dance, to capture the essence of the system the way a portrait artist captures the essence of their muse.

One of the explanations for the "one-hundred-thousand-year problem" is that the climate cycles we see aren't exactly one hundred thousand years in length, but in fact vary from around eighty thousand to one hundred and twenty thousand years. Since both of these numbers represent a rough multiple of the forty-one thousand year obliquity cycle, it has been proposed that something in the complex suite of climate feedbacks effectively amplifies the deglacial warming phase of every second or third obliquity cycle, leading to ice ages of varying lengths that average out to one hundred thousand years.[27] Other studies have gone a step further and proposed that astronomical forcing itself is almost irrelevant in pacing the ice ages. Instead, they suggest that it is the inherently chaotic nature of the climate, combined with feedbacks within the climate system, that leads to oscillating temperatures that self-organize to a state in which the periodic changes become fairly regular.[28] Most likely it is some combination of both astronomical forcing and chaotic internal variability, in which feedbacks within the Earth system control the climate within certain bounds[29]—a limit cycle like those we encountered previously, in the predator-prey species interactions of chapter 2.

To understand how a system can be externally driven by a smoothly varying quantity like the astronomical changes in solar radiation reaching Earth's surface and yet still exhibit chaotic fluctuations that are entirely unrelated to that forcing, we can turn to early mathematical experiments in population ecology. Robert May, or more formally, the Right Honorable Lord May of Oxford, was a theoretical physicist and professor of zoology who pioneered the mathematical exploration of ecology in the 1960s and 1970s. In particular, May was interested in whether two isolated but interacting species would eventually establish some sort of balance in which the predator would reduce the population of the prey species until it was too sparse to support the population of predators, which would die off and allow the prey to recover and start the cycle again. Using a simple equation, May showed that such a scenario was indeed possible under certain conditions. The equation May used has a variety of forms but is commonly referred to as a "logistic map." The logistic equation of population growth was first derived in a simpler form by the Belgian mathematician Pierre Verhulst in 1845, countering the theory of exponential population growth proposed by the English cleric Thomas Malthus in 1798. Instead of a population that grows at an ever-increasing rate, Verhulst argued that there would be a limit beyond which further growth would be restricted by the carrying capacity of the system—perhaps the availability of land or food, for example.[30] Logistic, rather than exponential, growth therefore gave rise to an S-shaped curve, also known as a sigmoid. This S-shaped curve initially starts with very slow growth before rapidly accelerating, just like an exponential curve. But as the system becomes increasingly constrained by its carrying capacity, the curve flattens out once more and further growth is impeded. It was through using a modified version of this scheme that May had demonstrated how, under conditions where a population consisted of predators and prey, the flat, top part of the curve might not always remain constant but would fluctuate through time. But he also showed, just as Lorenz had demonstrated with the atmosphere, that if the governing conditions were only slightly altered, the same equation could produce chaotic behavior in the top part of the S-curve, in which populations fluctuated wildly, even to the extent that one of the species would be completely wiped out.[31]

But what has this to do with the climate? Well, it turns out that the pattern of global temperature change through the ice age cycles of the last five hundred thousand years or so, as revealed by the ice core and marine sediment records discussed earlier, can be modeled as an inverted S-curve. That is, the slow change at the beginning of the cycle represents the stable, relatively warm conditions of the interglacial periods. Then, instead of rapid growth, our inverted S-curve reflects rapid cooling of the global climate as it plunges quickly, driven by accelerating feedback, into the ice age. And then the rate of cooling slows down. But just as May had found with his predator-prey system, this part of the curve was far from one of monotonic change. Instead, as the slow-but-steady cooling continued, the climate as a whole became increasingly vulnerable to chaotic behavior. At this "glacial maximum" point, far from the equilibrium of the interglacial climate, only a minor perturbation was needed to collapse the entire glacial climate system and drive rapid, self-sustaining, climatic warming. And that is precisely what we were witnessing in the abrupt atmospheric warming event we saw earlier in the Greenland ice core histories at the end of the last ice age, 11,500 years ago.

And it is also precisely because the system is so sensitive to any small perturbation when it is at its glacial maximum state that the differing climate model conclusions described previously might actually *all* be correct. The climate system isn't a train of toppling dominos, each tile falling neatly, in turn triggering the next, and then the next, and so on. The climate is a tower of wooden blocks, like Jenga. As the tower grows, it becomes ever taller but also more and more vulnerable. No player can predict which brick, when removed and placed at the top, will be the one that collapses the whole structure. And each time the tower is rebuilt, it may be a different block that proves to be the catalyst, even though the tower itself looks remarkably similar to the one built previously. Any system that is incrementally pushed further and further from equilibrium relies increasingly on a sustained and unwavering flow of energy in order to maintain its form. If that flow changes or if the energy that it provides fluctuates too much, the system is eventually pushed away from its steady state and into one of chaotic and unpredictable change. Gaia, enveloped by Uranus, is still not immune to the occasional storm.

In the case of the ice age cycles of the last half-million years or so, this oscillation between long glacial periods and shorter interglacial periods traces out a pattern of asymmetric fluctuation that repeats approximately every one hundred thousand years. *Long breath in, quickly out.* The long cold glacial portion starts with a rapid cooling that then slows but nonetheless continues to cool. The shorter interglacial portion starts with the very abrupt rise in global temperature and then a period of much more stable conditions before the cycle starts again. We have encountered this type of asymmetric behavior before, in the rainfall patterns of the early Earth in chapter 1, in the pattern of speciation and extinction in the fossil record described in chapter 2, and in the *S*-shaped curve of the logistic map described earlier in this chapter. These behaviors, and the pattern of global climate change through the ice age cycles since the Mid-Pleistocene Transition, can all be characterized as relaxation oscillators, the pattern of change that arises when a steadily changing quantity eventually reaches a critical point. From that juncture a rapid succession of feedbacks quickly releases the energy that built up slowly in the preceding phase of gradual growth. The release of this energy leads to large-scale restructuring of the system and a return to its initial state.

The Past Is the Key to the Future

The possibility that climate might behave chaotically, as Lorenz had discovered, and that it might switch abruptly from one state to another as the ice cores showed, presents us with a worrying outlook for the future. In the IPCC report, the idea of long-term "commitments" was well highlighted—the concept that, once we adopt a certain trajectory of change, we might be "locked in" to that trajectory for many more centuries, or even millennia, even if we reduce greenhouse gas emissions. The evidence supporting the idea of committed change comes both from climate models and from geological evidence of warmer-than-present periods of the past. But the geological evidence tends to mostly reflect the "end game" of a particular climate scenario, that is, the altered steady state that the past global environment achieved once all of the fast- and slow-acting feedbacks had fully played out. So to understand what the transition period might look like, as our climate shifts from one steady state to another, we have to rely on models. The trouble is that, as we saw earlier, general circulation models tend to smooth out the chaotic behavior of the global climate to remain computationally stable. Because of that, we might be grossly underestimating the turmoil we might endure as our climate flip-flops between differing states, rapidly switching between familiar and predictable conditions and those that might once have been considered "extreme."

Just as the Greenland ice cores revealed at the end of the last ice age, average temperatures in some parts of the world could reasonably shift several degrees Celsius in just a few years. Geographical variability would bring about distinctly different climate responses in different areas, so although uncharacteristic heat waves might even start affecting the Arctic or Antarctic, as they did in 2021 and 2022, other parts of the world might suffer flooding, wildfires, or anomalously cold winter weather. According to the International Disaster Database,* the average number of global weather mega-disasters (those costing at least US$20 billion) has increased threefold since 1990. The four mega-disasters of 2021 led to nearly a thousand deaths and collective damages estimated at US$150 billion.

In a landmark paper published in 2018, professor Will Steffen from Australian National University, together with a team of international experts, put forward a radical and alarming vision for humanity's future.[32] Titled "Trajectories of the Earth System in the Anthropocene," Steffen and colleagues used evidence from the past, together with insights from models and recently observed changes, to suggest that our current appetite for carbon-rich fuel sets us on a path away from the glacial-interglacial pattern of variability we have experienced for the last million years or so, and instead toward a "hothouse Earth," the likes of which haven't been witnessed for tens of millions of years. The authors suggest that "this pathway would be propelled by strong, intrinsic, biogeophysical feedbacks difficult to influence by human actions, a pathway that could not be reversed, steered, or substantially slowed." Indeed, the term "feedback" appears in their paper

---

* Launched in 1988 by the Center for Research on the Epidemiology of Disasters (CRED) with support from the World Health Organization (WHO) and the Belgian government, the Emergency Events Database (EM-DAT; www.emdat.be) catalogs the occurrence and effects of mass disasters around the world in order to assist with humanitarian aid. So far, they have collated data for twenty-two thousand such events, since 1990.

nearly sixty times, such is the importance of these processes of interaction that lead to cascading impacts from one component of the Earth system to another. Positive feedbacks abound. Warmer air in the Arctic thaws permafrost, which then releases into the atmosphere carbon dioxide from the organic matter previously frozen within it. Methane, a greenhouse gas that, over a one-hundred-year time frame, is thirty times more effective at trapping heat than carbon dioxide, is abundant in seabed sediments, locked up, under intense pressure, in an icelike solid form. As oceans warm, the cold conditions necessary for these hydrate deposits to remain "frozen" diminish, and methane escapes to the atmosphere. Warmer oceans also promote greater bacterial populations that respire oxygen and release carbon dioxide, and because warmer water freezes less often, the extent of polar sea ice declines. Without this extensive and highly reflective surface, the Earth absorbs more heat from the sun, further accelerating the loss of ice and amplifying atmospheric warming. But worse still, it is not just the abundance of positive climate feedbacks that is of concern. Due to human activity, some of the negative feedbacks that could help regulate our climate, such as carbon uptake by tropical rainforests or the absorption of carbon dioxide by seawater, are also already becoming less effective (in the former case because of deforestation and burning, and in the latter case because our oceans also absorb heat from the atmosphere, and gases tend to dissolve less readily in warmer water).

How then might we achieve a "stabilized" instead of "hothouse" Earth and avoid what some researchers consider, "a ghastly future"?[33] Clearly a better recognition of climate feedbacks is paramount. Understanding the relative import of each and the role of amplifying processes with which they might connect to form cascades of change that rapidly upend the comfortable steady state to which we have become accustomed is key. We must seek the counsel of a wisdom somewhat greater than our own. Our ancestral mother of all life has recorded for us clues, a user manual for the world we now recklessly inhabit. The climate feedbacks and the behaviors described earlier are all written in the archives of the Earth, encoded in rocks and mud and ice, patiently awaiting our attention and decryption. From the pages of these scriptures, we know that, for the last million or so years, our climate has absorbed energy from the sun and distributed or absorbed it in ways that led to remarkable consistency in pattern of change. A relaxation oscillator, slowly cooling, rapidly warming, respiring mindfully as it traced a path from stable state to something far from equilibrium and back again. Insignificantly small deviations in the amount of solar energy reaching our planet, periodically fluctuating over hundreds of thousands of years, were amplified on Earth by a system of coupled oscillators, each one breathing at its own rate and governed by its own rules. Collectively these coupled components transmitted change from one realm to another—atmosphere to ocean, ocean to ice sheets, ice sheets to land—and from pole to pole, continent to continent, in each case triggering impacts that cascaded through the global system as thresholds were breached and the functioning of individual components switched between alternate modes. During the last million years our climate has achieved a state of impressive organization, and although the insight of our proxies fades as we look back further, the evidence for such a well-regulated condition seems to suggest some uniqueness to this relatively recent accomplishment. Has our global climate, through myriad interactions and feedbacks with all the physical and biological elements of the evolving Earth system, gradually and incrementally adapted itself in ways that increasingly favored not just the evolution of life, but the emergence of complex life and self-aware, conscious, intelligence?

Progressive optimization of our global climate system since the tumultuous birth of our Earth 4.53 billion years ago has been possible because of its feedbacks. Forests of the Carboniferous period drew such copious amounts of carbon dioxide from the air that our climate cooled[34] and oxygen levels increased. And because that carbon was captured in lignin-dominated plants growing in swampy environments, it was unable to decay, eventually becoming locked up in rocks that are now three hundred million years old. Under the more favorable climatic and atmospheric conditions that followed, life flourished, diversifying and expanding into almost every available ecological niche on the planet. Wandering tectonic plates that swallowed oceans, built mountain ranges, gave birth to volcanic arcs, and ruptured Earth's crust with seismic activity all played a role in governing how our climate came to the state it was in when *Homo* inherited it from its Australopithicean predecessors two or three million years ago. By the time *Homo sapiens* emerged three hundred thousand years ago, our climate was comfortably oscillating within the limit cycle of glacial-interglacial variability, a breath of life rising and falling to a mantra of one-hundred-thousand-year solar forcing. Albedo feedbacks during these one-hundred-thousand-year cycles triggered the slow onset of each successive ice age and the slow cooling of the planet, forcing our ancestors to migrate with the animals they relied on for food. But eventually these ice age worlds became too dry, too dusty, and starved of energy by expansive ice sheets and sea ice that reflected light and heat back into space. And over just ten thousand years—quickly in geological terms—that frozen world collapsed as reinvigorated ocean currents brought heat to ice sheet margins left vulnerable by lowered sea levels. A chain reaction of positive feedbacks brought rapid changes that propagated quickly through the ocean and atmosphere, transforming the land surface as the oceans rose and flooded the coasts and liberating fresh new expanses of terrain for displaced flora and fauna to reclaim once more. Like the S-curve described before, the slow start of deglacial warming rapidly accelerated as feedbacks promoted ever-faster change, until ultimately the disequilibrium was vanquished, and the stability of our current interglacial climate was achieved.

And for the last ten thousand years, this climate has maintained near-constancy, fluctuating imperceptibly during this time by only a few tenths of a degree Celsius. Yet now, since our discovery of the tremendous power of that carbon-rich fuel fossilized beneath us, we have warmed the Earth more than one degree since the mid-nineteenth century. The Atman of our new, industrialized world belches carbon dioxide and methane into the atmosphere faster than our climate feedbacks can accommodate or adjust to it. *Humans* are indeed a unique species, but who are we, and how did we get here?

HUMANS

4

That's one small step for [a] man, one giant leap for mankind.

—COMMANDER NEIL ARMSTRONG, SETTING HUMANITY'S
FIRST FOOT ON THE LUNAR SURFACE, JULY 20, 1969

For twenty-eight hours in July 1969, Major General Michael Collins was 238,855 miles from Earth, entirely alone. During each one of *Columbia*'s thirty lunar orbits, Collins was invisible for forty-eight minutes, tracing a silent arc through the distant black vacuum of space on the far side of the moon. Relaxed and assured with his experience, Collins reported that he felt no fear or loneliness, just "awareness, anticipation, satisfaction, confidence, almost exultation."[1]

But Collins was not the first to have seen the dark side of the moon. One year earlier, in December 1968, Frank Borman, James Lovell, and William Anders flew the *Apollo 8* spacecraft around the moon ten times without landing. Escaping Earth's gravitational pull by accelerating to 24,200 miles per hour—around 7 miles per second—*Apollo 8* was the first time our species had left a low Earth orbit and reached deeper into the unknown emptiness of space. As with Collins, the crew were on their own, out of radio contact, for thirty-four minutes of every orbit. "We'll see you on the other side," said Lovell as *Apollo 8* prepared for the first radio blackout and its insertion into lunar orbit on the dark side of the moon. *See you on the other side*—a simple linguistic construction suffused with symbolic meaning. We say this to one another as we face hardships, even death. Yet to the astronauts of *Apollo 8*, it was meant literally, a nonchalant sign-off as they embarked on a feat never yet attempted by any other life-form that we know of.

And it was with this undiminishable evolutionary significance in mind that NASA manager Abe Silverstein had named the lunar program after the Greek god Apollo, a charioteer whose daily task was to move the sun through the heavens and in doing so bring light and life to Earth.* From the eighth century BCE, Apollo was one of the most favored gods among the Greeks, considered the most beautiful and wise. Apollo was the oracle at Delphi who offered prophesies and protection. He was the god of colonization and patron of seafarers, of music, poetry, and dance, of healing and disease. So it was that through his symbolic association with education, reason, and order that Apollo was also the guide who saw to it that boys became men and that the wisdom of the gods was shared with humankind.

Although Apollonian virtue was adored by the Greeks, they accepted also that god's darker brother, Dionysus, a representation of chaos and disorder, religious ecstasy and insanity. The coupling of these dueling deities—artistic Apollonian spectacle and the Dionysian personal destruction—found expression and meaning in Greek theater through the tragedies popular at that time. In *The Birth of Tragedy*, Friedrich Nietzsche conceived of two separate worlds, dream and intoxication, in which the former (Apollonian) world yields unto us "beautiful illusion" and the latter (Dionysian) world, "a complete forgetting of the self."[2] Through tragedy, the myths of the gods were made real on Earth, with all their beauty, violence, and despair. Nietzsche believed that with the decline of tragedy, so came the decline of the myth. But maybe our myths just changed. The Greeks looked out from Earth to the stars to create their legends; now, in the space age, we create new ones not just by looking out ever further into the void, but also by looking back down.

For our "coming of age" as a species, which enabled the safe escape from our planetary bounds, brought with it an entirely new perspective. During that first crewed exploration of the moon, *Apollo 8* astronaut William Anders took a photograph of the Earth, a distant crescent rising above the lunar horizon. "Earthrise" was called "the most influential environmental photograph ever taken" and considered by some to mark the beginning of the modern environmental movement. Poet Archibald MacLeish commented that "To see the Earth as it truly is, small and blue and beautiful in that eternal silence where it floats, is to see ourselves as riders on the Earth together."† This framing of what was primarily a technological feat in such emotive language speaks to us deeply, connects with

---

* www.nasa.gov/centers/glenn/about/history/apollo%20press%20release.html
† https://hls.harvard.edu/today/fantastic-voyage/

us fundamentally, and resonates with us honestly and humbly, just as the myth of a charioteer god pulling the sun through the heavens did for the Greeks. And for us, like them, the power of such a vision is that it connects us as individuals to one another and to the planet we call home.

Origins

Seven million years before we set foot on the moon, our ancestors separated from apes and eventually became the first humans. But this was not a singular, explosive, divinely orchestrated evolutionary thrust that propelled into existence a lineage of unique and specialized hominids. This gradual unzipping and divergence of genetic code was one of many such radiations that, as we saw in chapter 2, tentatively provoked the landscape of possibilities until it was sure that that land could bear its evolutionary weight. Our origin lies in a radiation that played out over eons during a period of time known as the Miocene. Around sixteen million years ago, during the Mid-Miocene Climatic Optimum (the warmest parts of that geological age), there were more than fifty kinds of apes exploiting diverse habitats over a wide geographic range. As the global climate shifted, slowly cooling and changing the habitats that they were accustomed to, selective pressures first split the ancestors of modern orangutans, then gorillas, then chimpanzees from our genetic line. By six or seven million years ago, the first true ancestors of modern humans were walking the Earth, standing more upright than their ancestral chimp cousins with whom we still share 98 percent of our genetic code. Interbreeding most likely continued for another million or so years, but eventually, by around five million years ago, we said goodbye to the other great ape lineages and strode out alone into the evolutionary wilderness. *See you on the other side.*

The culmination of four-and-a-half billion years of terrestrial, biological, and climatological organization, adaptation, and optimization brought forth, in the early Pliocene, a new ape that would forever transform our planet. Some suggest that this "coming of age" for the great apes was driven by the deteriorating climate and the "specialization trap." This trap arose from adaptive feedbacks that had initially allowed our ancestors to develop dietary, cognitive, and locomotor adaptations that were optimized for their previous environment but that now, as open savannah replaced forests, left them less able to develop new skills and strategies.[3] Against a backdrop of gradual adaptations, the slow and steady developmental changes that had been ongoing since the Miocene, new combinations of genetic traits and social behaviors started to emerge that shaped the species we have become today. Just as the *Apollo 11* astronauts learned quickly to adapt and optimize their movements when they first christened the untouched dust of the lunar surface with boots constructed a quarter of a million miles away, so too did Australopithecus adapt its behavior and embrace the possibilities of the increasingly unfamiliar world.

Escaping the trap of tree-living specialists wasn't easy. Primate body plans are typically more upright than most other mammals; we have a high degree of joint mobility and three-dimensional color vision, but to make the adjustment to a whole new way of life, a life lived in the open, low to the ground, exposed to predators, and intermittently far from shelter required something new, something unique. And that new skill was bipedalism.

All apes are capable of a bipedal stance, and some, such as gibbons, frequently run along branches on two legs. So the new skill of bipedalism was not one of anatomical change; it was one of evolutionary need. Primates have most likely employed two-legged climbing and running for the last six million years or so, but it persisted as something interchangeable with knuckle-walking and four-limbed climbing until a feedback exerted by the climate on the biosphere cascaded into a selective pressure that drove early hominids (apes and early humans) to fully embrace this new mode of locomotion. And over a couple of million years, this pressure gave rise to the specializations famously preserved for eternity, frozen in time, in the volcanic ash of Laetoli, in northern Tanzania. The Laetoli footprints are remarkable, an irrefutable snapshot of two-legged walking more than three-and-a-half million years old. Discovered in 1978, the canonical example of Australopithecine bipedalism established for the first time a relatable point in our history to which we could point and say, *we were here*, in precisely the same way that those boot prints on the moon will speak to the generations that come after us. Yet Laetoli gave up another less well-known secret. Two years before the discovery of those first human footprints, another set of prints was found, described to have a "rolling and probably slow-moving gait," a creature that "shambled" rather than walked, crossing its feet as it did so.[4] For decades these prints were thought to record passage of a bear, but in 2021 a new study reexamined the footprints with modern imaging techniques that allowed for a more complete and objective biomechanical analysis.[5] The results were astounding. Not only did the new analysis refute the idea that these tracks were made by a bear, but they argued persuasively for their formation by human feet. But they were not the same Australopithecine feet that formed the more famous set of prints; they were from an altogether different species. The conclusion is inescapable—multiple species of human lived in the same habitat at the same time, and both were walking upright. In contrast to those footprints on the moon, the evidence from Laetoli shows that, at least in the middle Pliocene, we were certainly not alone.

In the millennia that followed, our bipedal ancestors adopted new behaviors made possible by freeing up their hands. Digitally liberated and manually dexterous Australopithecus became an increasingly advanced hominin, progressively adapting to its changing landscape through cultural and behavioral optimizations. Sometime after three million years ago, a new species emerged, *Homo*, and from that point on, the pace of change accelerated exponentially. Like the phase transition we encountered in the Cambrian explosion of chapter 2, the rapid ascent of *Homo* from a close Australopithecine cousin 2.8 million years ago, to *Homo erectus* 1.7 million years ago, to *Homo sapiens* only two hundred thousand or three hundred thousand years ago was driven by environmental and climatic feedbacks that fueled anatomical, cognitive, social, and behavioral adaptations that optimized our species for the diversity of climates and landscapes we live in today. But when, why, and how did early *Homo* leave its ancestral home in the African plains?

Habitual fire use was common from around four hundred thousand years ago. Whether for cooking, warmth, or defense, harnessing our ability to maintain, transport, or make fire afforded us an adaptability and flexibility that earlier hominids hadn't had. In short, we had become more resilient. But for at least two hundred thousand years, our genus, *Homo*, remained geographically restricted to the African mainland. Populations across east Africa, from what is now Ethiopia, Kenya, Tanzania, and south to Zambia, Zimbabwe, and Botswana, thrived in areas where vast lakes punctuated the dry savannah plains. The Makgadikgadi-Okavango wetland was one such area, now much reduced by tectonic and climatic changes, that sustained diverse and plentiful wildlife and provided a favorable habitat for human occupation. This region played a critical role in the later development of modern humans and in the emergence of Middle Stone Age technology—a more sophisticated toolkit than that used by their Palaeolithic ancestors, showing greater refinement of tools and adaptations designed for specific purposes.[6] With a climate now locked in to regular but asymmetric one-hundred-thousand-year-long ice age cycles and wetter or dryer periods oscillating with the more frequent twenty-one-thousand-year cycles of Earth's wobble around its axis, innovation, adaptability, and resilience was the key to survival for early *Homo sapiens*. As the climate of the last half-million years or so, the Late Pleistocene, became cooler and drier, life became harder. Lakes contracted, vegetation became less abundant, less lush. Coastal populations started marine foraging,[7] but in the interior the surviving wetland oases felt increased pressure as their area reduced. For *Homo sapiens*, their gradual shift from the primarily plant-based diet of ancestral primates to the omnivorous foragers, scavengers, and hunters they were becoming presented both opportunity and danger.

As human toolkits improved, large-bodied grazing mammals declined. Today, subsistence hunters only target smaller prey when large prey are depleted—the energy expenditure increases as prey size declines, and the rewards lessen. In learning to hunt, we first removed the easy prey, and in doing so initiated a feedback loop that forced us to be more resourceful, more creative, to use our brain more than our brawn. Climatic change, as our planetary relationship to the sun waxed and waned over tens and hundreds of millennia, set the tempo of our species' viability, but it was our relationship with the wildlife around us that ultimately encouraged us to migrate.

The climatic, environmental, and cultural feedbacks that both pushed and pulled *Homo sapiens* to stretch its phylogenetic wings are so intimately intertwined that to assign one or another as the most important has limited value. But what most definitely *is* helpful is the quest to understand what might have been possible for those early migrants and the likely time line tracing out the challenges they faced. Computer models now guide our interpretation of the sparse and incomplete archaeological archive, revealing to us physically constrained reconstructions of statistically plausible scenarios, accelerated in time like the *Apollo* astronauts breaking free from Earth's pull. Spinning up the world in bits and bytes allows us to see those scenarios that match the data, the relics of lives lost, lovingly scraped, brushed, and dusted from the Earth. Discarding those with least agreement, we build up our portrait of Stone Age *Homo*, archaeologist becoming artist, meticulously teasing detail from pigmented layers to illuminate that which once was hidden.

And from this endeavor a now-familiar pattern comes into focus. As our climate warmed or cooled, as Earth drenched or dried, and as plants flourished or floundered, corridors of possibility opened and closed and beckoned or held back our inquisitive adventurers. Today there are no hunter-gather communities in areas that receive less than ninety millimeters of rainfall a year, no grazing animals, no vegetation. In the Middle Stone Age, windows of migrational opportunity were constrained by the physical environment just as they are now. So as the flickering switch of equatorial climate tipped one way and then the next, populations moved in pulses through northern Africa, crossing into Eurasia by way of the

Nile-Sinai land bridge where the African mainland joins Saudi Arabia or further south across the Strait of Bab-el-Mandeb, where the Red Sea is narrowest and could be crossed during times of low sea level.[8] Our history, the history of *Homo sapiens*, is thus one that bubbled up in the forests, plains, and wetlands of Africa, the crucible of humankind, and repeatedly experimented with ideas, possibilities, and combinations of contingences. Leaving their footprints in the mud of past worlds, our ancestors were telling us, "We were here, and we were not alone."

Although experts still debate when the earliest migrations began, how persistent or intermixed their populations might have been, or what driving forces motivated their inexorable colonization of nearly every accessible fragment of land on Earth's surface, physiologically we know that certain traits arose directly from the environmental conditions under which they thrived. During the last two million years, both brain and body of most *Homo* species have increased, and during at least the last million years or so, the changes in body size appear to be related to the gradual decline in average air temperature.[9] Rainfall, by comparison, the critical control on migration, seems to have had little effect on our growth, and neither temperature nor rainfall have substantially influenced our brain. Intriguingly, recent research has uncovered an even more direct environmental driver of behavior, one controlled specifically by our exposure to ultraviolet B (UVB), a key component of sunlight. Combining results from studies in mice as well as humans, researchers found that higher UVB exposure was associated with hormonal changes that led to enhanced male-female attractiveness and sexual responsiveness, and in men, higher testosterone levels and increased aggressiveness. But of greater surprise was that these effects were most apparent in men originating from countries with relatively low ultraviolet radiation, meaning that, as early *Homo* populations migrated steadily northward, their reproductive timing became more seasonally skewed toward late spring and early summer, just as it is in Europe today.[10]

Life Becomes Lifestyle

Environmental drivers exerted pressure on *Homo*, but it was us who closed the feedback loop by responding to these pressures in ways that modified the opportunities available for our future evolution. By escaping the "specialization trap" of tree-dwelling primates, humans became free to roam diverse habitats and migrate as food and climate allowed. But the increasingly rapid ascent of humankind through the hierarchy of life on Earth still needed something extra, something more than just free rein over an unspoiled world.

Controlled use of fire, as we saw earlier, allowed early humans protection from cold and from predators and, through cooking, extended both the spectrum and energy availability of edible foods.[11] As tool use increased—whether stone, bone, or wood—fire allowed materials to be more easily worked and modified. Using fire to alter landscapes gave humans a much more rapid mechanism to engineer their ecosystem and to transform their environment from one that they simply encountered to one that they chose to inhabit. And increased habitation was likely one of the consequences of routine fire use, for it required concerted effort to collect fuel and a central place to which it could be brought. Through fire, humans developed a new requirement for group cooperation that complemented the social structures necessary for communal hunting. And through these two developments emerged not just a life, but a lifestyle, a way of living that brought individuals together for a common cause. Fire lengthened the days and brought extended families together for longer periods, shaping social interactions. It is tempting to speculate as to the role fire played in the origins of myth, legend, spirituality, and the sense of wonder that often comes over us as we stare into the hypnotizing dance of a flame. Yet there is something else about the emergence and adoption of fire use that reveals a profound transition in the origin story of early humans. Fire was not just a tool like a hand ax, a technology that spread slowly from Africa into Europe over many hundreds of thousands of years as waves of migrating hominins took the toolkit of their time with them and passed on stone-shaping skills to their kin. Fire use appeared quickly, in widely distributed parts of the populated world, and in different subpopulations within the greater corpus of *Homo*. This difference between the early spread of hand ax technology and of the later use of fire suggests that people no longer needed to physically take skills with them in order for them to disperse. By now, it was *ideas* that people were spreading, innovations that had no physical solidity but that were instead primarily cognitive advances.[12] Such a shift in behavior has far-reaching implications, for it necessitates repeated and meaningful interactions between disparate social networks, ones that were nonviolent and cooperative, even among nonfamily groups. In short, the habitual use of fire and the interactions through which it so rapidly spread marked a turning point in human evolution, and that turning point was the origin of culture.

This diffusion of ideas appears not just in fire use, but in the spatially uneven and temporally nondirectional emergence of other culturally significant practices, such as engraving, the manufacture of ornaments, and the fabrication of bone tools specifically for fur and leather working.[13] Continually driven to innovate by changing climatic and environmental conditions, Stone Age groups continually explored novel creative ventures and, when possible, shared these with those they interacted with. Together these social and physical factors gave rise to a pattern of human evolution shaped both by climate and by cultural connectedness.[14]

Connecting social groups allowed the diffusion of ideas and the establishment, in the last few hundred thousand years, of a rich and diverse human cultural landscape. And in turn, those interactions and the skills required to facilitate them fed back on the cognitive abilities and physiological evolution of those early migrants. Studies in modern humans from countries across the world show that, as a species, we are best able to navigate in unknown territories if they are like the environments in which we grew up.[15] More significantly, it seems that these associations tend to persist throughout our lives, suggesting that experience shapes brain structure early in life, and that regardless of age, sex, or educational ability we are, to some extent, navigationally optimized for the types of landscape with which we are most familiar. When we're on our home turf, we feel most secure, most capable, and most connected to our immediate social network. But remove us from that bubble and we struggle: life becomes harder; it demands greater cognitive investment and physical energy expenditure. In fruit flies, chronic social isolation induces significant loss of sleep. It also leads to metabolic changes that confuse the brain, signaling starvation and triggering overeating. In humans, as well as flies, persistent social isolation brings about new emotional states that become increasingly intense over time.[16]

So physical and social proximity to one another naturally emerge as critical parameters for humans to thrive. We are social creatures, evolved from group-living primates whose intimacy with each other defines and is defined by a social hierarchy. But unlike other primates, who typically only recognize maternal, and not paternal, kinship, humans tend to form so-called thick relationships with others regardless of sex. Thick relationships have strong attachments to one another, a sense of obligation and of mutual responsiveness.[17] How nonverbal infants infer which relationships are "thick" and which are not has, for a long time, remained unclear. This is not surprising, given the limited possibilities for infant communication in either direction. New research, however, has uncovered that toddlers and young children rely on saliva sharing to determine whom to trust and to define which of the many people they are continually but transiently surrounded by are family members and not just friends or caregivers. Kissing, as well as sharing food or common eating utensils, enables infants to quickly identify their kin and to rapidly build a conceptual understanding of "family." That this mechanism emerges early in life, with no explicit teaching, suggests that we are hardwired to place greatest emotional attachment on those of our group who share an intimate (saliva-sharing) connection with us, because those are the people who we believe will respond to us in times of distress and who will protect us above all others. This intimacy feedback self-strengthens over time but is plastic as relationships come and go or as social structures are redefined. Through kissing we implicitly confer kinship on one another, family status that extends beyond shared genetics and allows our sense of security and attachment to be continuously updated and refined.

For a child, the maternal connection can be so strong that sometimes the two become one, a behavioral and physiologically connected unit in which coordinated actions—responding to one another's cues—modifies the parasympathetic nervous system in ways that synchronize their heart rate, cardiac activity rising as engagement with the child increases.[18] Behavioral and physiological synchrony through instinctive feedbacks driven by one another's cues therefore plays a critical role in our healthy socioemotional development as children and shape the adult humans we go on to be.

For growing infants, a critical stage of development occurs when they master the art of speech. From that point on, their relationship with the world changes forever, as they no longer simply rely on empathetic responses of a proximal parent; speech enables children to command and question any number of bystanders with a richness and nuance that grows rapidly with practice and is optimized based on success or effectiveness of each new interaction. And so it was for our ancestors, yet the earliest of them had no language to learn; for them, the vocal landscape was unknown, untested, and full of possibilities.

Not all language is spoken. We can write, sign, or whistle, and even without language, vocalizations themselves can impart a range of meanings depending on how they are delivered—we innately perceive the difference between a warning shout and a soothing murmur. So language can be multimodal and modality independent; we have several ways to communicate and that makes language entirely distinct from other types of animal communication. Theories on the evolution of speech are numerous, but as a purely physical process, speech required nothing more than modifications to the primate vocal apparatus. We vibrate our vocal cords, open or close our nasal chamber, and shape the sounds with our lips. But no other ape uses its tongue to modify sounds. In English, the word "language" comes from the Latin "lingua," meaning "tongue," reflecting the importance of this crucial speech articulator that not only allows us to speak, but to shape sounds that can be faithfully copied and repeated. Because to be useful for communication, sounds must be consistent and have a specifically assigned meaning. How certain sounds became attached to their respective meanings is widely debated; we simply don't know. But there are some theories that make intuitive sense and show similarity with processes we've already encountered earlier in this book.

One particularly intriguing possibility is that speech was a self-organizing process that arose from interpersonal responses—a feedback in which imitation of one person by another spontaneously gives rise to a repertoire of sounds shared by an entire group. As babies we first learn to babble, making syllabic sounds like "da-da" that allow us to gradually gain control over speech articulators. Humans and bat pups are some of the only mammals to indulge in babbling,[19] yet it seems that this stage might be fundamentally important for infant speech development. Through babbling, and later as we begin to make more controlled and repeatable sounds, an adaptive feedback loop develops that ensures that with each interaction, communicators are able to constantly check, adjust, or update their sound library to stay linguistically in sync with their audience. One interesting aspect to this idea is that speakers and listeners have different preferences and so play distinct roles in this self-organizing process. Speakers tend to prefer sounds that are easy to pronounce and so prioritize articulation over clarity. By contrast, a listener prefers sounds that are clear and distinct, even if those sounds are difficult to make. As people in a conversation naturally take turns at both roles, speaker and listener, a linguistic tussle ensues that eventually resolves to some sort of compromise in which both ease of production and clarity of signal are optimized.

Physiological adaptation combined with increased need for reliable social interaction and sharing of complex ideas undoubtedly drove the emergence of language, as opposed to more simple vocalizations. But it was language's synergy with the neural pathways associated with tool use that really accelerated this fundamental transition in human culture. Although a link between good syntactic language and tool use skills had been discovered a few years ago, it wasn't until 2021 that new research found direct evidence that the fine motor skills typically employed when using tools activated the same region of the brain that helps us understand the syntax of complex sentences. The basal ganglia sit below the cortex of the brain, the outer layer, and are not only associated with control of voluntary body movements but also with learning, cognition, and emotion. That this area should be heavily involved in both tool use and language skills is therefore not particularly surprising. But what is remarkable about this shared neural resource is that training in one skill leads to improvement in the other, suggesting that the shared neural network is primed by one activity in a way that facilitates greater ability in the other activity.[20] With such a tightly interwoven learning framework it is easy to see how a positive social feedback might arise, in which skill-sharing and teacher-learner interactions within a group actually made the evolution of language easier, which in turn accelerated learning and social interaction, more effective sharing and learning, and so on.

Incredibly, this feedback didn't just serve to improve skill sets in language and tool use, but it also helped us hone the predictive processing strategy that our brains use to parse external information and assign meaning and importance to it. Our brain doesn't work by simply making a "best guess" as to how incoming sensory signals should be interpreted. It makes predictions and then evaluates those predictions against the available information. By taking into account the ever-changing relationships between words in a sentence, for example, predictive processing allows the brain to much more quickly and accurately respond to a stimulus. This way of processing cascades through levels of comprehension, from the low-level prediction of upcoming sounds, right through to accurate prediction of individual words.[21] Even more remarkably, our brain remains plastic, or reprogrammable, throughout our life. Congenitally blind people have been taught to "see" by using the visual cortex to interpret sounds in the same way that bats and dolphins use complex sounds to infer geometric shape, suggesting that different regions of the brain are primarily wired for specific computations, specific predictions, and not for specific senses.[22]

Prediction, as we explore in chapters 5 and 6, plays a crucial role in our lives at all levels, from the individual to the social group, and forms the basis of the adaptive process of feedback, which helps to incrementally optimize a system. And the way we predict words, or communicative elements, is in part influenced by the environment in which we live and the relative importance of the words or concepts to which we are exposed. A simple example is the way that we talk about color. A human with good visual acuity can perceive a far more extensive color space than they can easily describe in words, which means that we have to compress our visual perceptions into an approximation that can be used for easy communication. But this process of approximation varies among individuals, depending on the salience of the color being described.[23]

Landscapes differ in the statistics of color occurrence—in some environments some colors are more abundant than others. But depending on our lifestyles we might also find that some colors are simply more important than others. For foragers it might be more important to differentiate between multiple shades of red, for example, to convey greater detail about the ripening state of gathered fruit.

Our predictions are the basis for subconscious decision making, which in turn influences our ability to thrive and survive as individuals and as a species. In the example above we saw how our linguistic repertoire evolves in tandem with our predictive model in ways that are governed by our external environment. But it isn't just the words we hear that affect our predictions. How they are spoken can also play a significant role in shaping the inferences we make. As a species we choose mates based on a range of qualities, of which voice pitch is one component. To males, a female voice with a higher pitch is perceived to be more feminine and younger, perhaps enabling mating males to identify females closer to their most fertile years.[24] Extending this further into the modality of music in which pitch variation is intricately combined with tonal and rhythmic structures, we discover that our brains respond to music differently than to other sounds. There are distinct populations of neurons in the brain that respond to music and speech, but even more specifically, there are areas of the brain that respond almost exclusively to song.[25] Since singing or listening to music with singing can evoke emotional responses and trigger memories through interactions of the neural circuits responsible for each, it leaves us to wonder how the evolution of speech, then language, then song may have imbued our emerging Stone Age culture with meaning, empathy, and a wider sense of connectedness.

Running, Awakening, and Learning to Learn

To the nineteenth-century philospher Nietzsche, Greek tragedy emerged directly from "the spirit of music," an association that recognized the "astounding significance" of song. Music, Nietzsche argued, "forces us to see more and more deeply than we otherwise would"; it is "the actual idea of the world," a magical and essential thing that invigorates and energizes, brings life and insight and clarity. For animist cultures such as the Indigenous Australians, song was a part of their creation myth; it was the way that the legendary beings brought the world into existence. Walking the "songlines," the "footprints of the ancestors," awakened every bird, animal, every plant, rock, or water hole.[26] This invisible labyrinth of meandering pathways that crisscross Australia might be tens of thousands of years old, a visibly hidden but tangible and significant feature of Aboriginal Australian culture that connected people to their landscape through music, storytelling, painting, and dance.

Our capacity for these more sophisticated mental talents—language, abstract thought, self-awareness—relies on the thick outer layer of our brain, the cerebral cortex. This layer accommodates only sixteen billion of the eighty-six billion neurons in a human brain, but the neuron density of primate cortices is far greater than that of other mammals. Somehow, somewhere along our evolutionary trajectory, we found a way to radically increase the thinking capacity of our brain in precisely the right region to allow all the necessary skills for cultural—not just physical—development. As it turns out, the neuron density of the cortex follows the same kind of power-law relationship we have discovered in earlier chapters, the self-similar scaling relationship that maintains its form regardless of size and which appears to characterize a multitude of natural systems.

Daily energy use in humans also follows a power-law relationship, in this case related to fat-free body mass.[27] Adolescents utilize nearly 50 percent more energy than adults, and then, as we age beyond sixty, our energy expenditure drops, and we lose muscle. But it isn't just muscle that is using all that energy. Despite amounting to only 2 percent of our total body mass, our brains use a disproportionately large amount of energy, as much as 20 percent of our caloric consumption at rest. Feeding this voracious organ required us to prioritize glucose transport for cognition over athleticism, such that compared to our closest relative, the chimpanzees, our gene controlling brain glucose supply is three times more active and appears to have undergone far more mutations than is likely by chance. Strength, it turns out, isn't just a function of muscle mass; it arises from the careful neurological redirection of energy—a directed adaptive process that rapidly increases our apparent strength even before new muscle growth has taken place. As we see later in this chapter, evidence is emerging elsewhere that accelerated and even *directed* genetic evolution seems increasingly plausible, even if we don't yet fully understand how or why. But regardless of the causal mechanism, our evolutionary history reflects adaptations that have tended to optimize our neural functioning for enhanced cognitive—and especially cultural—performance. Perhaps through singing, dancing, and staring into campfires, we gradually awakened our inner self.

Our awakened brain, our cognitive and cultural development, all relied on processes that interacted with other processes, feedbacks among systems that slowly but significantly optimized our physiology and behavior in ways that made us better adapted to our environment—one that now included diverse species of coexisting humans all vying for a larger share of evolutionary longevity. To drive this adaptive cycle of incremental optimization, we needed energy, and lots of it. Energy density came to direct our daily lives, focusing our activity on strategies that satisfied our burgeoning energetic demands while simultaneously maintaining an efficiency of acquisition that still allowed us to live the culturally rich and socially interactive lifestyles we were so rapidly developing.

Bipedalism emerged early, as we saw earlier in this chapter, but it was only when environmental pressures forced us onto the open grasslands that we preferentially adopted upright walking. The more we walked, the better our limbs adjusted to this mode of locomotion, and the easier it became for us to exploit new habitats and sources of food. As we walked, we became more successful, and the more successful we became, the more we chose to walk rather than climb. This positive feedback self-selected not just for upright locomotion but for more rapid migration and more wide-ranging social contact. Our legs have evolved to incorporate collagen-rich tendons and ligaments that store up elastic energy as they are stretched and then release it to propel our body forward. But this mechanism isn't used during walking, and compared to most other four-limbed animals, humans are poor sprinters. So it seems that as we episodically and perhaps furtively first explored the opening landscapes around us, we learned to run, not fast, but far.[28] In terms of metabolic cost, our optimal walking speed is a little more than one meter per second. But by exploiting the spring mechanism in our legs, we can move at two-and-a-half times that speed, and with a lower energetic cost. Running favored mouth breathing to maximize airflow and dissipate heat, and we lost body hair to increase convective heat loss from our skin. Most likely we adopted endurance running around two million years ago as a way to access high-energy food sources such as meat and bone marrow, which first appear in the archaeological record a little earlier. Endurance running isn't seen in any other primate, most of which are primarily plant eaters, suggesting that our demand for energy to fuel our rapidly growing brain played a formative role in our behavioral development and evolutionary success.

Our brains grew rapidly from around two million to one-and-a-half million years ago and demanded from us a lifestyle that delivered considerably more fuel. We ran to hunt, but the myth that "meat made us human"—perhaps by enabling bigger brains—no longer stands up to scrutiny. In a recent reanalysis of the archaeological record, there was no detectable increase in the evidence of carnivory among early Homo,[29] so something else was driving brain growth. Compared to other primates, human subsistence strategies tend to be more intense, requiring a greater investment of effort but providing greater energetic rewards. This mode of "intense and behaviorally sophisticated subsistence" yielded more fuel for the time spent, but required greater social cooperation and food sharing among family groups.[30] But the reward was more leisure time, more time to interact with one another, more time to learn from older generations, and, who knows, perhaps more time to sing.

Dirt

Living off the land imparted to us a vast population of invisible helpers, microbes from the soil that took up residence in every nook and cranny of our hairless, naked bodies. Thirty-nine trillion microbes—eukaryotes, archaea, bacteria, viruses, and fungi—working together as a "superorganism" in and on our bodies, communicating with the soil in a constant exchange of genes, growth-enabling molecules, and inoculants.[31] Our close contact with the land through the plants and animals that it hosted allowed our gut microbiome—this microscopic galaxy of assistants—to evolve in tandem with the microbiome of the soil. In the ground, this microbiome maintains the health of the soil for the higher organisms that depend on it, buffering them from stressors such as water or mineral loss, protecting them from harmful invaders. And just as the soil microbiome confers resilience upon the host it inhabits so too, through a conversation of chemical communication with our organs and physiological systems, does the microbiome of our gut.[32] Microbially produced enzymes digest our food, liberating nutrients, and synthesize vitamins we couldn't otherwise make, and by regulating our immune system and manufacturing anti-inflammatory compounds, our trillion-member team of helpers shaped our evolutionary success. Invisibly entangled, the health of our bodies depended on that of the living earth, and as we moved and migrated, our microbial colonies were replenished and updated as we absorbed from the land more and more new biota, enabling us to adapt to unfamiliar foods more rapidly than we could otherwise do. As bacteria in the land changed those in our stomachs, we found ourselves attracted to new sustenance and lost cravings for the foods we'd left behind. The plasticity of our microbiome allowed us to be wanderers, explorers, adventurers, and opportunists, while its ecological memory imbued us with a metabolic toolkit for dietary diversity and nutritional flexibility.[33]

The subsistence lifestyle, combining hunting when needed and foraging when possible, diversified our portfolio of energy supply and afforded us a resilience that promoted species success. But it also gave us something else, something so quintessentially human that, without it, Neil Armstrong would never have uttered those immortal words as he, for the first time in the known history of the cosmos, became the first person to set foot on a world other than his own. That new invention was science. Like the origins of language we saw earlier, subsistence relied on an ability to predict the future. If *Homo* needed large amounts of energy to satisfy the increasing energetic cost of its expanding cortex, it would benefit considerably from being able to acquire that fuel from the most energy-dense sources available. Gathering low-density fruits might have been a dependable fallback, with lower risks than hunting wild animals, but by gradually adapting to eat meat, our ancestors could secure much larger caloric returns for their daily efforts. To be successful, however, they would need to know where the herds would be. Tracking could be a simple affair: just follow the footprints until you find their owner. But this is an inefficient approach and prone to failure. Understanding what animal had left those footprints, its typical daily routine, need for water, shade, or rest, would allow a more intelligent hunter to speculate as to where the animal might be heading, perhaps even predict its movements far enough in advance that the hunter could lie in wait. Contemporary indigenous trackers share collective knowledge and critique each other's assumptions;[34] they use inductive and deductive reasoning based on the variety of evidence available to formulate hypotheses that they then test in systematic and carefully controlled ways. And not only that, but theirs was a creative science, one born of necessity and fueled by imagination and critical inquiry. They developed art as a vital and equal counterpart to their scientific methods so that their social communication and shared learning was as rich in content as it could possibly be. Art played a sufficiently important role both in survival and cultural development that it persisted and evolved, diffusing rapidly through social groups and from one population to another. Art revealed empathy in our awakened brains, allowed us to relate to animals of the hunt as if we were one of them. Tracking was an art, and that art was the origin of science.

For hunter-gatherers today, endurance running is not simply a means to an end, a way to explore their homeland in the search for food. It is intrinsically a spiritual and transcendental experience, one that connects them with the land not just physically, but consciously, purposefully. And so it was, too, for the fledgling humans that first unfolded their fragile evolutionary wings and felt the lift and supporting updraft of physiological adaptation. Those erstwhile pioneers of a new mode of locomotion soared across the opening landscape not just with a hunger for new foods, but with an unquenchable thirst for knowledge. From nature they learned, and through learning became gnostics, scientists, artists, and explorers. They were each their own Apollo, god of music, protection, reason, and education. The chariot they pulled each day across the open sky was one of increasing wisdom that they shared and let spread, person to person, family to family, tribe to tribe. Awakening the natural world, they awakened themselves, and through art recorded for others the rapture of their existence and the connections they felt with their kin. Connectedness, empathy, compassion, and a realization of their "not alone-ness" gave life to deeper neural pathways and pushed the electrical fabric of their expanding cortex to a new level of communication and comprehension. And as they spread their intellectual wings still further, they evolved skills for language and abstract thought, learned to run and dance and sing, and as they shared the evening warmth of their campfires with others just like them, they found they had a culture, and our species had at last come of age.

We Are Icarus

Through a vagrancy of metered migrations, our expanding Stone Age population was enticed not just east to west along climatic bands, but also northward, pulled and pushed by climatic flirtations that episodically opened verdant highways and virginal land bridges, a beguiling temptress that promised the world and the possibilities it contained. We were already a cultured species, but now, as we packed our growing bodies and flourishing tribes into ever more diverse and specialized habitats, we came to face a new challenge—we had to learn to live together.

In chapter 5 we explore the feedbacks that underpinned the emergence and cohesion of our societies, from the Neolithic to today, and how they shaped who we are as people. But for now we must finish our evolutionary discourse, because the origin story of humankind is but one part of *Homo sapiens*, the Apollonian part to which we must now present the unruly Dionysiac brother. He was always there, of course, as light demands dark simply to be, so too did the awakening early *Homo* feel the wrath and fury of nature, the chaos and torment that shattered Nietzsche's "beautiful illusion." But two million years ago, we took only that which helped us survive. Two hundred thousand years ago, we still left behind more than we took. Twenty thousand and even two thousand years ago, we still trod lightly upon the Earth, compared to today. What changed? Some blame the agricultural revolution, the last ten thousand or so years during which we, as a species, have progressively sought to control an ever-greater portion of global natural resources.[35] But when did growing our own food turn into denying other species their food? At what point did *Homo sapiens* decide that our survival no longer relied on nature, on natural diversity, on hierarchical and scale-free relationships like the Sheldon spectrum we encountered in chapter 2? When did we stop being a part of nature?

Perhaps in the last 250 years, the brief flicker of time in which our burning of fossil fuels has irreversibly changed the climate, driving our inexhaustible quest for progress, technology, commerce, and growth. In chapter 3 we saw how the Industrial Revolution forever shifted the pace and demands of our lives. But there are also more direct impacts, some that only now are coming to light. During the twentieth century, lead was a common additive to many products, such as paints and gasoline, and even before then, its softness allowed it to be easily worked and used for things like pipes. But lead is a neurotoxin, one that disrupts healthy development of our brain, bones, and cardiovascular system. Lead poisoning is insidious, invisible, and accumulates in the body over time. It is particularly harmful to children. A recent American study estimated that more than 170 million Americans alive today, more than half of the current population, were exposed to harmful levels of lead during early childhood. Millions of these experienced lead levels five or more times higher than the standard reference level.* What this means is that half of Americans have been neurologically impaired by lead that they absorbed from their environment, particularly from particulates in vehicle exhaust fumes. And this measurable impairment amounts to a drop in cognitive ability of 2.6 IQ points per person, for a total of more than 824 million IQ points, across the entire population.[36] Although subtle, these impacts most likely affected not just cognitive abilities but motor skills and emotional regulation as well. With the rapid phasing out of leaded gasoline in the United States during the 1980s and its eventual ban in 1996, the impacts will decrease over time, but those born from 1950 to 1980 will still be affected, perhaps imperceptibly, for the rest of their lives. And then there is the rest of the world, in which the United Nations Children's Fund estimates around eight hundred million children are still exposed to high levels of lead due to poorly regulated lead-emitting industries such as battery recycling.

Since the Industrial Revolution, we've increasingly added chemicals to our environment in pursuit of economic growth and technological prow-

---

* The "blood lead reference value" defined by the Centers for Disease Control and Prevention is not a health- or toxicity-based value; it is simply defined by the 97.5th percentile of the blood lead distribution in US children ages one to five years (see www.cdc.gov/nceh/lead/news/cdc-updates -blood-lead-reference-value.html).

ess. Lead has made us less intelligent, but recent research has shown that other metals released into our air and water are now altering the ratio of male to female births. In a combined study of American and Swedish human sex ratio at birth, lead in soil was found to correspond with a greater number of girls, but aluminum, chromium, mercury, and polychlorinated biphenyls (PCBs) were all linked to a greater proportion of boys. Intriguingly, at conception, the ratios tended to be indistinguishable from 1:1, suggesting that whatever influence these toxins have, it relates to embryo loss during gestation that then manifests in a skewed sex ratio at birth. Female embryos tend to be lost early, typically in the first trimester and early in the second, whereas male losses occurred late in the second trimester and onward. Whatever the causal mechanisms, it seems entirely possible that a small initial difference in sex ratio will cascade through hereditary feedback loops into a larger, more persistent disruption to the sex ratio of the wider population.[37] Combine this with observations that urban environments high in air pollution, road traffic, and road noise levels tend to cluster with higher values of body mass index (BMI) and higher rates of obesity, and it becomes alarmingly clear that in our desperate push for advancement, we are tangibly reshaping our demographics, our physical health, and mental acuity. Inside our bodies, the gut microbiome that once was so intimately entwined with that of the physical earth on which we trod is now evolutionarily mismatched with our environment.[38] As a consequence of a diet drenched in processed carbohydrates, protein, and fat, devastatingly devoid of fibrous plants and the natural antibiotic resistance they used to provide us,[39] our species now contends with a burgeoning epidemic of autoimmune disease and morbidity as well as the socioeconomic impacts they create.

These impacts feed forward into future generations, driving adaptations that align us with our increasingly unnatural (urban) environment in ways that are severely detrimental to our species' longevity and success. We are no longer Apollo, sharing knowledge, valuing reason and beauty, smoothing the transition from boy to man as our bodies and minds mature. We are instead Icarus, the boy from Greek legend who flew too close to the sun despite his father's warnings, only to fall from the sky as his waxen wings melted, plunging fatally into the sea below.

Evolution, then, is not just something that happened in the past or that happens so slowly we can ignore it. It is an ongoing adaptive process of optimization that perennially redefines who we are as a species by changing who we are as individuals. Thankfully, not all such change is bad. During the last 150 years, the Dutch have grown in stature more than three times faster than Americans, to become some of the tallest people in the world. Partly this could be due to favorable environmental conditions, but it is also evident that a positive feedback was at play. Across three decades, from 1935 to 1967, taller women had higher rates of child survival, meaning that subsequent generations were taller and so on.[40] Whether there is any evolutionary advantage to this remains unclear, but human females do tend to prefer taller males,[41] so from the perspective of an individual's genetic persistence, being tall clearly gives us a leg up on the evolutionary ladder. Other selective pressures have more obvious benefits, such as survivability in an environment that wants to kill you. Malaria, a blood disease carried by mosquitoes, kills more than a million people in Africa every year. There is a human genetic mutation that protects against malaria, but mutations have long been thought to be random, unpredictable. Those that lead to adaptations favoring reproductive success will survive; those that don't will die out. But recent evidence has uncovered something quite unexpected. The rate at which the malaria-protecting genetic mutation occurs seems to be higher in people from Africa, where malaria is rife, compared to in Europeans, where malaria is rare. In this instance, it is not that the "useful" gene is being preserved; it is that the ostensibly random mutations that create this gene are taking place much more frequently in the countries where those mutations are most beneficial. The authors of the study suggest that this is the first evidence of "long-term directionality" in human evolution and that specific environmental pressures can, under certain circumstances, actually influence how a person's genome evolves.[42]

Homo Futuris

The expanding global population of our species threatens all life on Earth, including our own. Current species extinction rates are nearly two thousand times higher than they were when *Homo sapiens* first came into being. Correlating extinctions with climate, particularly for mammals, shows no greater relationship than would be expected by chance. But correlating extinctions with human population size yields a predictive accuracy of 96 percent, leading some researchers to suggest that by 2100, all areas of the world will be experiencing a new wave of species extinctions[43] as our human population continues to rise. If global temperatures also continue their trend, humans themselves will be in jeopardy in large parts of the world. Human survival through humid heat waves exceeding 35 degrees Celsius is limited to around six hours, because of the physiological and thermodynamic limitations on our ability to cool ourselves down. Vulnerable members of our population, such as the elderly or those with chronic health issues, will be susceptible at temperatures well below this. And all members of society in already hot countries, especially those in economically disadvantaged areas, will be far more affected than those of us in more affluent or more temperate regions of the world. We are Icarus, and we are melting.

In 1996, the economist Robin Hanson put forward the idea of a Great Filter as a partial resolution to the Fermi paradox. In essence, the problem posed by the Fermi paradox is this: If extraterrestrial life is as likely as some estimates suggest, then why don't we know about it? With innumerable planets orbiting an uncountable number of stars scattered through myriad galaxies across immeasurable space, is it realistic to think that we, *Homo sapiens*, are the only form of advanced intelligence? The Great Filter concept helps rationalize this quandary. It posits that, even if intelligence did emerge elsewhere in the universe, for us to be able to detect it now, it would need to have survived any number of contingent events that could have led to its annihilation. Just as the dinosaurs went extinct and made way for mammals, so too could we go extinct, and some other order of life take over. Would they be more or less intelligent than us? Given what we know of our species' tendency to decimate others, perhaps the greatest chances of survival come from staying in the galactic shadows, avoiding other life that might be out there, just in case it combines superior power with the same bloodlust we appear to have developed. The Great Filter may consist of events in our evolutionary past—tests that we have passed in order to be here now. It may also encompass events in our future—nuclear war, climate breakdown, and ensuing catastrophic social unrest. Either way, for the evolution of *Homo* to continue, we need to once again connect with the Apollonian and dispense with the Dionysian.

Earlier we equated increasing technology and advancement with our increasing separation from nature. But done well, there is potential for technology to be our savior. The Kardashev scale measures how technologically advanced a civilization is based on how much of the available energy they are able to use. At present, we use about 1×10^{13} watts.[44] To get a foot on the first rung of the Kardashev ladder, we need to increase this by at least three orders of magnitude. But here's the catch—we need to find a way to harness one thousand times more energy than we presently do, while also staying within the ecological or planetary bounds that dictate our survival. To achieve this, we would need not only the rapid transition to a "cultural society" that values its ecosystem over economic interests, but also the rapid escalation of technologies that could harness far more of the solar energy than we already currently receive. Becoming a Kardashev "Type I civilization" would mark us as an advanced form of life, but to become a Type II "stellar civilization"—one that would enable us to spread throughout our solar system—we would need to capture at least one quarter (1×10^{26} watts) of the total available solar output (4×10^{26} watts).[45] A Type III "galactic civilization" requires yet another ten orders of magnitude greater energy. Although such esoteric imaginings might seem like little more than light entertainment, they highlight something genuinely important. In our evolutionary quest to be successful, to proliferate and survive, to spread our genetic code as far and wide as possible, we have to balance technological advancement with careful stewardship of our environment. Because without the latter, no amount of technology will save us.

Will we, as a species, ever fully "come of age" and find a way to live in balance with all other life while also enabling the pursuit of our innate desire to learn and to grow? Through our evolutionary ascent, starting seven million years ago in the forests and grasslands of Africa, we have adapted to the world around us, shaped our behaviors to optimize our functioning for the cyclically changing climate and the variety and inconstancy of the habitats it laid out for us. Feedbacks turned existing skills such as intermittent bipedalism into a characteristic trait and then a strength that enabled us to roam further and faster, gathering more energy-dense foods that fed our swelling brains. Our larger brains enabled more meaningful interactions with one another, which in turn led to new skills, learning, and intelligence that propelled our brains yet further along their steepening trajectory of growth. Through deepening and increasingly complex neural structures, we evolved reasoning and scientific thought. We discovered art and empathy, humility and connection; we found song and transcendence, culture and spirituality. In a universe thermodynamically destined to evolve in the direction of increasing entropy, chaos, disorder—the Dionysiac—we have found ways to build ideas from information and to build ideas into order and complexity.[46] Adaptive evolution, the key to our survival, relies on our ability to predict our environment with accuracy and across timescales. We are a learning system, and such systems fare best when the variables of change are separated in time—some fast, some slow. To make such predictions, we must observe and adapt. Life did not emerge by chance in a finely tuned universe. It rose up, purposefully, in a universe that was self-tuned for its existence.[47]

Where next, for this purposeful evolution? "We choose to go to the moon," wrote Ted Sorensen for President John F. Kennedy's address at Rice University, Texas. It was September 12, 1962, and the *Apollo* missions to the moon were still an ambitious dream. "We choose to go to the moon in this decade and do the other things, not because they are easy, but because they are hard."[48] What other species chooses to do things "because they are hard"? It continues: "because that goal will serve to organize and measure the best of our energies and skills." Purposeful action, meticulously planned, carefully executed.

Seven million years and six crewed moon landings since we came down from the trees, we are a species who knows where we have come from, knows where we need to go, but doesn't yet know how to get there. As the popular saying goes, "if you want to go fast, go alone; if you want to go far, go together." We are not alone: we have each other; we have culture, we have *Society*. *We'll see you on the other side.*

SOCIETY   5

No man ever steps in the same river twice, for it's not the same river and he's not the same man.

—HERACLITUS

Five thousand years ago, a unique and monumental episode in human cultural evolution began. Across the British Isles, from southern moorlands to the islands of the far north, Neolithic families worked together to design and engineer flat earthen platforms a hundred meters or more wide and to construct upon them circular arrangements of colossal standing stones. We will never know the words they used with each other, the language that gave them cohesion and commonality. And we might never know for certain what underlying force united them with such purpose and commitment. But what we do know is that, over the millennium that followed, coherent, organized, and socially well-connected groups built a thousand or more such structures, each adorning an upland or otherwise prominent seat in the landscape, using materials distantly quarried and deliberately imported, and employing labor from far and wide.

Why would they embark upon such a struggle? What reward could they have envisioned to justify the sacrifices they voluntarily made? For those who built them, these magnificent works of art had meaning and purpose, their slow and deliberate creation a bursting forth of a collective intelligence that had evolved gradually through the millennia before. In building stone circles these Neolithic groups proclaimed to the world that they valued their future and chose to shape it to their advantage. Because each structure brought order to the natural world, set in stone the cyclic changes that watermarked their daily lives and that repeated dependably from year to year, one generation to the next. Predicting the ebb and flow of seasonal changes, the circles these settlers constructed became the basis of a contract with their environment. With the help of majestic stone markers, they could each year be better prepared, more prosperous and prolific than the year gone by. The short-lived struggle of construction locked in long-lasting benefits and adaptations that eased their toil and enabled growth. Through such monuments, these people harnessed the incredible power of feedbacks.

This period was one of purposeful and far-reaching change, a turning point that marked the establishment of a farming culture in these far corners of northwest Europe, a society that settled instead of roamed, cultivated instead of gathered. Technological changes had migrated slowly north from the developments of the earliest civilizations in the Middle East another five thousand years before them. It was a time when regional climates were ameliorating, imperceptibly changing as the steepening tilt of Earth's axis brought warmer summers to northern Europe, and the last vestiges of the great North American Laurentide ice sheet decayed and released its grip on the overturning circulation of the North Atlantic Ocean. In northern Scotland, a remote peat bog archives the comparative dryness and milder temperatures of this climatic optimum, each millimeter or two of sediment recording the conditions of a single year.[1] Eight thousand years of history, hidden in mud, buried beneath the surface of an unknown and isolated mire. And this mud can tell stories, can recount to us now what those early people could not. From bromine levels we know how rough were the seas, and from the clarity of liquid extractions we can reconstruct how the climate shifted from wet to dry. As Neolithic villagers cut, hauled, and thrust upward their rough-hewn slabs of Orkney sandstone, we know they did so in a place that was less battered by the wild Atlantic storms than it is today, where each successive year saw increasingly favorable weather, and one where the living was attractive enough to justify the perilous crossing from the Scottish mainland.

Whether favorable climatic conditions were instrumental in the establishment and proliferation of henges and stone circles or whether their contemporaneity was purely coincidental, the investment of time, energy, and resources necessary for an undertaking as substantial as any of those whose lithic skeletons we still see today suggests an importance and purpose that has led many to infer spiritual motivations. We have no knowledge of the religious beliefs of these people, but the extant evidence attests to the considerable technological skill of this society. The Neolithic village of Skara Brae, on the west coast of the Orkney mainland, is the most complete example in Europe. Occupied from 5,200 to 4,500 years ago, the ten houses in this settlement had stone hearths, beds, and recessed "cupboards" in their walls. The houses incorporated basic toilets, each connected to a simple, common sewer system. Each house followed a similar layout with the same furnishings throughout, including stone storage boxes made watertight with clay, a dresser, and seats. "Low roads" connected the village to henges and stone circles across the island, and to Maeshowe, a chambered cairn and passage grave seven miles to the southeast of Skara Brae, most likely built around 4,800 years ago. Here again, the implemented technology was purposeful. A thirty-five-meter wide and seven-meter-high earthen mound enshrouds a complex of stone-walled passages and chambers, whose constituent blocks weigh up to thirty tons. Together with the labor effort required for construction, estimated to have been in the range forty thousand to one hundred thousand person-hours, the evidence indicates a considerable workforce and many months, if not years, of work. And this work most certainly must have been carefully planned, controlled, and coordinated, for the chief distinction of Maeshowe is that, at sunset each winter solstice, the sun's rays are so perfectly aligned with the eleven-meter-long, low-roofed and narrow stone-walled entrance passageway that only at this singular point in the rotation of the Earth around the sun is the central chamber completely illuminated, just briefly, before the sun slips once more beneath the horizon to mark the end of the solar year.

Movement, Migration, and Connection

Neolithic structures aligned with the solstices enabled those early farmers to track the passing of time and to calculate precisely when at last the long dark winter nights would start to release their stranglehold on both light and life. Marking the solstices allowed those early societies to plan more deliberately for the year ahead, to predict instead of react. Social cohesion enabled these monumental constructions, and through the predictability they conferred, these monuments supported further growth and success of their makers. This simple feedback was perhaps part of what justified such labors, and these monuments survive to this day because of the scale and solidity of their construction, and because they have been revered and intentionally preserved ever since. Mathematical analyses have attempted to decode all manner of geometric relationships, yet the most fundamental observations to be drawn from such works is that each of these titanic crystalline theaters reflects the monumental struggle of a people who wanted to see and be seen. "We are here" proclaimed their monuments, and through solstitial alignments, that concept of "here" was placed within a larger cosmological framework. Their ancestors were most likely part of a wave of migrants that energetically babbled forth from the well-spring of the Levant, far to the south in what is now the Middle East. And as this tide first trickled, then surged, whirled and streamed across these northern lands, those migrants brought tools and techniques that became ever more refined, more innovative, and more purposeful. Migration mixed tribes, knowledge, and ideas in a maelstrom of learning and adaptation, and whatever the beliefs of the European settlers, they committed themselves to a shared purpose that offered security and prosperity, but at the same time demanded from them individual hardship and sacrifice. Building chambered cairns, passage graves, and wide henge platforms adorned with stone circles all necessitated colossal and coordinated effort to dig enclosing ditches through solid rock, to quarry giant blocks of stone and raise them aloft to rest recumbent on equally giant uprights, all made stable by the deep holes into which they were sunk. Whatever force it was that drove these engineers, it was tremendously unifying and motivating, and sufficiently important that, in some instances, they were prepared to dismantle these enormous creations and move them, stone by stone, across land and water.

Stonehenge, the most famous of all Neolithic stone circles, has graced the Salisbury Plains of southwest England for five thousand years. It is an incredible work of art, of science, of human ingenuity and social strength. But the so-called bluestones of Stonehenge were quarried not from the English plains, but from the Welsh hills, far to the west. And now, new findings show that these pillar stones were not freshly cut, but were recycled. Stonehenge wasn't an entirely new edifice; it was a relocation of a stone circle older still, one constructed among the Preseli Hills of west Wales at a site called Waun Mawn.[2] One of the earliest stone circles in Britain, Waun Mawn was built sometime between 5,600 and 5,000 years ago. Its diameter of 110 meters matches that of the enclosing ditch at Stonehenge almost perfectly, with an orientation to the sunrise of the summer solstice, just as we see today at Stonehenge. But the stone placement at Waun Mawn was less regular than Stonehenge, and although some pillars were left behind, others, such as the Altar Stone, were brought to Stonehenge from yet another source. So this colossal Neolithic theater was one that, over centuries or millennia, was adapted, refined, and rebuilt more carefully and deliberately with materials from not just one, but multiple distant quarries. What of the quarrymen whose labors it was that raised up these stone temples and breathed life into their form? Strontium analyses of human remains at Stonehenge uncover a story not just of unimaginable determination to move an earlier constructed stone circle 150 miles to a new location, but one that also reveals that those interred at this most re-vered of Neolithic sites included many who lived, until their last years, in the Welsh hills from where those stones came.

Coordinated social groups undertaking inconceivably difficult tasks thus migrated with apparent ease across the landscape, bringing with them livestock, technology, and culture. Flowing freely across the emerging so-cial riverbed of ideas, Neolithic societies pulsed into new lands and wetted the fertile soil of human ingenuity with ideas admixed with those from continental Europe. All those thousands of years ago, the early popula-tions of what became Britain and Europe, long divorced by rising seas, were nonetheless still intimately connected, sharing techniques for pottery, metalwork, and ritual burial. This "Bronze Age," an epithet borrowed from the historic stages of Greek mythology, was a time of innovation and long-range social connection. But gradually, in the centuries of the newly emerging "Iron Age," those connections faltered or were lost. Migration into Britain declined, and the invigorating tide of social learning began to ebb away.[3] According to Greek legend it was the flood of Deucalian that brought this era to an end, unleashed on humankind by the god Zeus. Sometimes wrathful, Zeus was the god of thunder, the head of the Olym-pian gods and the supreme deity of the Greeks. Yet Zeus was also consid-ered benevolent and just, a peacemaker who sought to resolve conflict and bring order to the lives of gods and mortals alike.

The balancing of these competing forces, like Apollo and Dionysus encountered in chapter 4, set the template that shaped those early societ-ies. In chapter 4 we saw how social order and connection underpinned early human evolution, enabling cooperation, trade, and the emergence of culture and shared values. From the social networks that arose in the civilizations that eventually sprang from those populations emerged a col-lective intelligence whose strength was in its structure, in the scaffolding that continually separated and recombined as it too was washed with the competing forces of contingent catastrophe and the determined drive of slow but steady progress. The story of society therefore is not one of linear nor regular evolution through time, it is one whose details are so complex and interrelated that, to understand how and why we became the groups we are today, we need to tease insights from different times and different places, compare and contrast the forces that challenged us and those that lent us resilience. The story of our societies is one we must piece together from the scattered debris of history, left strewn across the global landscape when the floodwaters of change have finally slipped away.

In chapter 4 we discovered the "songlines" of the Australian Aboriginal people, who sang into existence the world around them. Fifty thousand years ago, long before the quarrymen of the Neolithic, the earliest settlers of the Sahul* inhabited a land so vast and open to exploration that the possibilities for learning were immense. Encompassing tropical rainforests, temperate uplands, and hyper-arid deserts, this newborn continent was baptized by the arrival and spread of substantial populations. Over just a few thousand years, these new pioneers moved far inland, discovering sources of water and traveling between them, navigating first by learning and then by recognizing the prominent features in the landscape around them. Memorizing what they learned, they sang their land into existence, passing down to each new generation the songs that encoded what was known, what needed to be known. Across a continent four thousand kilometers wide, nomadic groups established superhighways that computer analyses now reveal to be the most "attractive" paths—the routes that connected water sources most directly and that benefited most fully from prominent navigational markers.[4] But equally important for a society trying to survive in a harsh and unforgiving landscape, these songlines, these transcontinental superhighways, followed paths that were the most energetically efficient. Like water that seeks the easiest channel, we flooded the land intelligently and iteratively, each time placing our feet carefully, learning, refining, and optimizing our path.

Elsewhere in the world at this time, our emerging societies also established trade and exchange networks across hundreds, even thousands, of miles. In eastern and southern Africa, groups traded ostrich eggshell beads, seashells, or obsidian among patchily distributed populations.[5] Climate forces controlled our mixing, but local hubs lent resilience in times of environmental uncertainty, times when long-range connections were weakened or collapsed, as mighty rivers such as the Zambezi began to flood more frequently and extensively, making migration routes impassible. But just as with the later Neolithic henge builders, migrating and mixing African societies traded something more essential for their cultural evolution than just new tools or techniques. It was something our earliest ancestors had first discovered as they stood upright and learned to run

---

* The Sahul was the continent of Australia and New Guinea, once connected by lower sea levels during the ice ages.

and hunt and outwit their prey. Through trade and exchange, geographically and culturally disparate societies built shared values and kept tokens that reminded them of those agreements. The "shared realities" we constructed gave a platform for mutual respect among individuals and lent cohesion to our groups.[6] And these shared realities enabled ideas to travel faster than the bodies they came from and, in doing so, formed the basis of a social feedback that accelerated our ability to learn, to adapt, and to optimize our collective behavior.

Wrath of the Gods

Yet even as our societies became ever more able to make predictions that could assist in population survival, there was still the threat of the "unpredictable"—the climatic switching that could, in a geological instant, trigger sweeping cascades of environmental change that stole from us the certainty of understanding that we once had and the efficacy of the tools we had so carefully built. The decline of the Indus valley (Harappan) civilization three-and-a-half thousand years ago has been tied to a change in the South Asian monsoon at this time that brought less rainfall to the rivers along which this great civilization was built. After two thousand years of success, all the cultural and technological progress that had been achieved was undone, in part, by the rain.

Elsewhere we benefited, instead of suffered, from the vagaries of our constantly evolving gaseous envelope. Civilization as we know it might not even have come to exist had the mid-Pleistocene climate not brought wind fields that, around two hundred thousand years ago, allowed silt from the Sinai-Negev sand dunes to be blown northeast into the Negev Desert.[7] The coarser material newly deposited in this corner of the Levant allowed fertile soils to develop that, at the time, contrasted markedly with the thin and barren soils that were more prevalent elsewhere. The more productive Negev soil layer allowed agriculture to take hold and in turn fueled the rapid emergence of the complex societies we know today. The hand of all-powerful Zeus meted out adversity then reprieve, balancing the scales of social justice impartially and unpredictably.

Today we might envisage that our abundant modern technologies provide us with more than sufficient information to understand and interpret our world, to make predictions for the future, and to adjust our behaviors accordingly. But for our historical and prehistorical ancestors, there was scant information and untested ways of knowing. If wind-blown silt could give rise to the birth of modern civilizations, and monsoonal changes be the downfall of at least one of those, what other contingent events might be lurking in our history, muddying the waters of our evolutionary narrative and raising timid questions as to our social foundations and institutional resilience? Some have argued that climate chaos triggered by one of the largest volcanic eruptions of the past twenty-five hundred years, from the Okmok volcano in Alaska, set in motion a series of environmentally driven socioeconomic feedbacks that sufficiently destabilized the Roman Republic that, within a few years, a new Roman Empire emerged.[8] Yet throughout its five hundred years, the Roman Republic was in an almost constant state of war,* so tracing a causal mechanism for societal collapse is fraught with difficulty and an ambiguity that some researchers suggest prevents such clear-cut inferences.[9] Certainly the historical records are far too incomplete to establish a train of events sufficiently clearly that unequivocal conclusions might be reached. But the evidence of a major volcanic eruption in the closing decades of the first millennium BCE, which triggered both cooler and wetter climatic conditions, perhaps leading to famine and disease, can at least speak plausibly to the idea of an external stress that was, perhaps, the final straw for an already collapsing society. If the volcano had triggered successive years of failed harvests, it most likely weakened social cohesion to such an extent that the demise of the republic was all but inevitable.

Coarse graining, seeing the pattern but not the details, affords us only a glimpse of the past, one that offers insights, but not necessarily answers. To understand specifically *how* our social structures fare when faced with adversity—how, why, when, and where they fail—we must reconstruct those histories more finely. To do that we might turn to more recent historical events, ones for which abundant and detailed written accounts survive. From these accounts we might be able to better tease apart the nature of the catastrophic forces at work, the scale and direction of the metaphorical flood they triggered, and the circumstances that together constituted the timbers of the social ark that allowed our forebears to survive. In seventeenth-century Finland, three volcanic eruptions (1600, 1640–1641, and 1695 AD) cooled the local climate by approximately 2 degrees Celsius as volcanic clouds injected particulate sulfur into the stratosphere. Cooling was relatively short lived, yet grain harvests and socioeconomic disruption persisted for many more years. With failed crops, farmers had no seed for subsequent years, so they deserted their homes and migrated elsewhere. But the feedbacks that arose from each successive eruption varied from one

---

* www.britannica.com/place/Roman-Republic

year to the next and from one region to another, because the differences in social infrastructure began to play an increasing role as each region was differentially stressed. Regional impacts depended on the type of agriculture that was practiced, the availability of material capital and resources, or the strength of infrastructural and social networks.[10] So although the external environmental pressures were common, the social and individual responses were not. Intervention by the state helped some, by relieving tax burdens on peasant landowners, for example, but by excluding the landless population from this assistance, epidemic disease quickly spread through the poor. And because diseases care not for status, it then spread to the wealthy. Social standing, therefore, didn't always guarantee survivorship.

Our emerging societies were, as we still are, at the mercy of Mother Nature. But as we increasingly adopted a more settled agricultural lifestyle, we began to benefit from a more positive feedback—one that still challenged us, but that also allowed us to adapt. The social turning point that marked the beginning of the Neolithic had brought greater population densities and closer human-animal contact. And in the millennia that followed, the number of pathogens that we became exposed to soared. In response, our bodies were evolutionarily selected for immunity against intracellular pathogens and for inflammatory responses to extracellular microbes.[11] Disease and famine continually stressed our societies, but through physiological optimization, social network connectivity, and behavioral adaptation, we faced each new threat better prepared than the last, endured each successive flood with a vessel more comprehensively equipped and more skillfully navigated. We rode out the tantrums of Zeus and hung on, just long enough, so that the "father of all" might have mercy and let us rebuild.

Under Pressure

So far, we've seen how societies shaped and reshaped themselves as environmental forces impacted and challenged them. And through these challenges, the groups that survived were the ones who had found resilience by adapting to the conditions and by making changes that would help navigate similar events in the future. In that way, societal evolution has a lot in common with many of the feedback loops we have encountered in earlier chapters. Often these feedbacks arose as responses to an external forcing, a "drive" that pushed a system to a state in which it continued to respond by adjusting its components one by one until the resultant adaptations were able to more easily absorb and dissipate that external flow of previously disruptive energy. We saw, for example, how the cooling outer shell of the early Earth fractured in ways that allowed tectonic plates to move and, by doing so, absorbed and dissipated energy from upwelling mantle plumes and enabled the continual recycling of Earth's surface. The life that then took hold evolved in ways that produced a scale-free distribution of global biomass, again driven by external energy that was absorbed by organisms of all sizes in ways that led them to self-organize, each size class being neither too numerous nor too few to maintain overall balance. Photosynthetic plants permanently altered our climate and allowed oxygen-breathing life to gain dominance over the anaerobes of the early Earth, and through the Carboniferous period, vast forests locked away carbon dioxide from the atmosphere and gave rise to a climate more favorable for species diversity. Our ancestors descended from the trees in response to the external drive of a changing climate, but in adapting to a more open landscape, we became more confident bipedal walkers and runners. The physiological changes that ensued allowed our species to become even more successful, developing language and culture, art and science. Cyclic behavior in our solid Earth, in our climate, and in speciation and extinction have allowed the same processes to be repeated, over and over, but each time different enough to enable new possibilities, learning, and adaptation. Land, life, and the energetic process of living all shaped and steered our ark to where we are now through a complex suite of feedbacks comprising innumerable actors and the infinite connections between them. But in each of these dynamical systems, the way in which their evolution proceeded was not just shaped by the external pressures acting upon them. Over time, they also felt pressures from within, from the competing preferences and demands of the myriad components of their complex, interconnected system. What can we learn from our past, then, to illuminate our history and shed light on how our societies fare when the driver of change comes not from outside but from within?

Just as the monsoon rains shaped the course of the three-thousand-year-old Harappan civilization in what is now the country of Pakistan, so too did the rains, five hundred years ago, set in motion a devastating sequence of events that forever changed the Maya of Central America. In the millennia before, Mayan kingdoms rose and fell as political allegiances continually reshaped the landscape. Theirs was a civilization that relied heavily on maize, a crop highly sensitive to drought during its flowering phase and unsuitable for long-term storage. In the first decades of the fifteenth century, drought conditions took hold and reduced agricultural productivity, disrupted long-distance trade with central Mexico, and led to food shortages that fueled population decline and migration.[12] Political rivalry, as leaders sought to gain advantage over weakened neighbors, escalated factional tensions, and widespread civil conflict ensued. A heavy reliance on a single food crop combined with a lack of centralized long-term storage and the competing authority of elite families produced a social system that was tightly coupled to the environment and so was vulnerable to external shocks, favoring survival only of the most resilient Mayan communities. And even those only held out for another century, until Europeans "discovered" the Americas and progressively conquered the fragmented Mayan territories. The fate of the Maya echoes that of the three ancient kingdoms of the Korean Peninsula a thousand years before, yet the weather-driven conflicts that arose there took place against a backdrop of political stability, producing "opportunistic" aggression in which affected states were more likely to be invaded rather than to attack their neighbors.[13] As with the Maya, the motivating mechanism for conflict in the Korean Peninsula was food insecurity, driven by weather shocks that preferentially affected those with lowest social resilience. How each community coped with those external shocks affected each of the neighboring groups, setting up a complex web of internal feedbacks that had the power to radically alter the course of history.

Long-term fluctuations in global climate correlate well with aggregate measures of war frequency and population change through a causal chain that links food shortage to price inflation, civil unrest and the outbreak of war, then to famine, death, and population decline.[14] But we are an adaptable and intelligent species, so the linear narrative of social disruption from direct climatic forcing is, perhaps, too simplistic. Instead, we might perhaps see climatic pressures as forces that narrow, widen, or redistribute the possible range of actions we might adopt.[15] Environmental changes force us to refocus our energy expenditure, and to be resilient, we might opt for innovation and creativity and to see opportunities in essential change. Resilience comes from how well we are able to capitalize on what is presented, how agile we might be as individuals and institutions, how we can transform our way of life to respond positively to the rising tide. "We are the flood, and we are the ark."[16]

In chapter 3 we learned about the International Disaster Database, the global compilation of mega-disasters resulting from extreme weather events. The data show such disasters are increasingly common and increasingly costly. The social and fiscal costs of damaging events are not just related to physical damage, but are also intrinsically linked to the ways in which each discrete socioeconomic "impact" is connected to each of the others. These connections might compound and accumulate "shocks" to a point where a regime shift takes place, and the socioeconomic system moves, perhaps irreversibly, to a new and different state.[17] Connections between societies, their infrastructures, and their economies all diversify risk, but they can also lead to far-reaching consequences from what were initially just localized events. In September 2017, Hurricane Maria made landfall on the Caribbean island of Dominica as a Category 5 storm. In the previous twenty-four hours, Maria's wind speeds had doubled from 85 miles per hour (Category 1 hurricane) to 165 miles per hour (Category 5 hurricane). As the storm tracked northwestward, it reached its maximum intensity, with wind speeds of 175 miles per hour. The tenth most violent Atlantic hurricane in recorded history made landfall in Puerto Rico shortly thereafter, having weakened, mercifully, to Category 4 (155 mile-per-hour winds). The wrath of Zeus that was unleashed in this epic tempest devastated a country already weakened by Hurricane Irma two weeks before, one whose failing electrical and water supply infrastructure was unable to cope. The power grid was completely destroyed, and 95 percent of cell phone networks were knocked out. Three thousand people lost their lives, and financial costs reached US$90 billion. But what of the long-term impacts? In the months that followed, researchers harvested anonymized cell phone location data, social media posts, and census and airline information, and reconstructed from the digital debris of a storm long past the ethereal footprints of a displaced and fractured population.[18] The smaller and more remote municipalities suffered the most in terms of the proportion of inhabitants lost or displaced and in the scale of damage to their infrastructure. With 80 percent of the country's agriculture wiped out, people moved from rural to urban centers. Others fled for Florida, Georgia, or New York. Puerto Rico suffered colossal damage to its people, national infrastructure, economy, and population, but during the months that followed, cascading effects were also felt across the United States. Prior to Maria, Puerto Rico had been a major supplier of medical resources, an industry that made up nearly a third of its economy. With many factories destroyed by Maria, shortages in medical exports to the United States led to depleted stock in US hospitals. In early 2018, coinciding with a winter influenza outbreak, these shortages left healthcare providers and patients thousands of miles from Puerto Rico struggling to cope.

What we learn from such catastrophes is that social inequity exacerbates disaster impacts, just as the historical records from areas as diverse as Central America and Finland showed us earlier. Unequal access to capital or resources renders the population differentially vulnerable to a uniform pressure, with marginal communities bearing the largest proportionate impacts both at the time of the disaster and in the months or years of subsequent recovery. The way we survive depends, in part, on our social capital. Social capital is a way of measuring the ability of individuals to benefit from their network of connections or wider group associations, and it is thought to play a critical role in resilience at all scales, from the family level, to neighborhood, city, and national or even global.[19] Shocked societies are weakened, even traumatized, and tend to relinquish freedoms, resources, or capital they would otherwise fiercely defend.[20] In such situations, greater social capital can mean greater social strength through a stronger cohesion of individuals. And this cohesion is an aggregate of many small-scale behaviors that change as we learn to adapt. Small learning loops, in which we learn by responding to events as they occur, help us navigate through a crisis. But larger learning cycles arise when we organize our resources and systems of interaction to build a more resilient social infrastructure that will cope more ably with future events. Combining these two components, we establish community-level learning and the self-organization of a group that allows adaptation and refinement. Restructuring our social organism in this way optimizes its operation for the changing external pressures. But there is something else that must underpin community learning, a solid foundation on which that new strength is to be built. And that something, that foundation, is social memory.

Lest We Forget

The weather gods of our civilizations—Zeus for the Greeks, Thor for the European pagans, Indra in Hindu mythology—are powerful and wrathful entities, able and willing to bring devastation to impious mortals. The thundering skies imparted such fear and foreboding in our earlier societies that even today we retain vestiges of these legendary gods in place names or days of the week: *Thursday* (Thor's day) in English or *Donnerstag* ("thunder day") in German. No matter how strong we became as individuals, families, tribes, or societies, we were never able to control the weather. But just as Neolithic henge builders learned to track the sun to predict the timing of seasonal change, so too have we learned to absorb and dissipate the continual bombardment of environmental forces whose violence and unpredictability remain outside our command. We learn to ride the wave rather than direct it. But those forces from beyond our realm are not the only challenges meted out upon us. Sometimes the gods of thunder, whether Slavic, Aztec, or Hindu, are also the gods of war.

Earlier we explored the climatic pressures and environmental shocks that triggered cascading impacts ultimately culminating in civil unrest and the outbreak of war. But war has many triggers, so what might we learn from societal responses to conflict? We might take a coarse-graining approach once more and identify commonalities among the chaos. War undoes development and social cohesion, regardless of the initial cause, through the physical and symbolic destruction of the fabric that binds us to one another, through the weakening of political institutions and by the fragmentation of our societies as the values, norms, and shared realities we once believed in are irrevocably shattered. In a theater of lost trust, how is the stage redressed and its actors reengaged?

Evidence collected in the aftermath of conflicts from El Salvador to Uganda, from Burundi to Nepal appears to show that, somewhat against the odds, social trust can, and does, recover. Communities once devastated by their own compatriots emerge from the wreckage more socially minded than ever, more willing to dedicate their time and energy for the greater good, keen to reconstruct and rebuild both physically and emotionally.[21] And this sentiment persists. A study of prosocial behavior among Vietnamese showed that those who survived the war became more deeply engaged in social organizations and preferentially participated in public affairs more than two decades after the conflict. Just as with the responses to natural hazards seen earlier, this persistence arose from two intertwined social responses. The first is an individual's growth after trauma, the "small learning loop" we saw before, which relies on a revision of the beliefs and assumptions once held as true. We seek to make sense of the world, to understand the cause and effect. Our prior ability to gauge one another, to predict each other's behavior, is lost, and a new set of values needs to be constructed, tested, and accepted. We need to believe that next time, we will be better able to "see it coming." To aid in this personal healing, we rely, too, on a second adaptation—the resuscitation of the larger social organism of which we were once part. Our connections, the fabric that intertwines us with one another and lends us strength, must be woven anew: a patchwork of families, blown apart, sewn back together to reassert control over the multitude of events that affect our daily lives. Shared grief and a shared sense of injustice are the messages we transmit to those who were not there, the young who we raise when the bombing has stopped, the children to whom we must explain why parents, siblings, or friends were lost. Participation in a new social narrative, built on the new values of those lucky enough to survive, brings comfort and identity, a social memory, lest we forget.

But what happens if we are pushed too hard, if the social memory of endured atrocities is so deeply scarred that community trust can never fully recover? For 350 years the Spanish Inquisition persecuted tens of thousands, perhaps even hundreds of thousands, of "heretics"—those individuals whose dedication to Catholicism after conversion from other faiths was deemed to be insufficient. Trials relied on the testimonies of acquaintances, friends, and even family members of the accused, an insidious and brutally effective mechanism that ensured a complete breakdown of interpersonal trust and a forced reliance on the state. The Inquisition held particular suspicion for the educated and more prosperous members of society—those whose learnings were perceived as a threat to the authority of the power-hungry church. Three centuries of persecution, etched into the land, the people, the economy. Using information from nearly seventy thousand Inquisition trials, researchers have revealed the lingering damage that can still be traced in the socioeconomic metrics of modern-day Spanish municipalities.[22] Where persecution was strongest, economic performance now suffers, educational attainment is lower, and trust is diminished. These are not populations who found solidarity in shared injustice; these are communities whose grief continues to trap them in a state of low education and low income, where local tensions run high, social trust is low, and the social capital needed to recover still hangs, delicately, in the balance.

Us and Them

In building our early societies, we harnessed feedbacks that ensured progress, and with larger populations came role diversity and a division of labor, a greater creative force that yielded innovations to further enhance our success. Yet negative feedbacks such as war and persecution reduced our numbers and dissolved the invisible ties between us, forcing us to rebuild what was lost, relearn how to trust, reconnect with those who became "others." And all this takes time. Trust must be earned; fears must be assuaged. As individuals, we are attracted to those we perceive as more similar to ourselves, and as we unite into groups, we continue to favor those within, rather than outside, our group.[23] And we do this to make things easier for ourselves, because there is a lower cognitive cost in staying close to those who we feel will behave and respond as we do, those that we feel we can predict. We draw mental maps of our social network and use those maps to navigate the social landscape. As our groups grow larger and as their diversity increases, mixing with others begins to yield increasingly positive outcomes. We adapt to the new environment of overlapping groups by sensing opportunities for future connections and shared social identity. Individuals in diverse groups paradoxically perceive themselves as more similar to each other, whereas those in more demographically homogenous groups retain views grounded in stronger stereotypes—they hang on to the "us and them." Increased social diversity is known to correlate with well-being, so how might we accelerate this at the community level—how can we hack the social system?

In a surprising set of experiments in Hungary, researchers explored how the feelings of individuals toward marginalized groups shifted as they changed their mode of interaction. Pairs of volunteers, one from the "in-group" and one from the "out-group," were asked to walk laps of a large room together, some pairs matching their strides to one another, others not. Those who adopted synchronized walking reported a greater empathy for their partners, a stronger interpersonal connection and reduced prejudice. Previous research on this kind of "motoric synchronicity" has concluded similarly, and though the causal mechanism behind this phenomenon might be complex, it seems likely that, by moving in time with one another, we blur the mental boundaries of "us" and "them"; *they* become part of *us*, part of our group.[24] And it doesn't need to be walking— any coordinated behavior can trigger this mental shift, even if maintained for only a few minutes. Even more remarkably, we don't even need to be physically present, simply visualizing coordinated behavior can better connect you both to those you know well and those you do not. No doubt the sense of unity between military personnel marching together arises at least in part from this kind of group-level motoric synchronization, a dissolution of the self and the emergence of a larger, more impersonal group identity. Interpersonal coupling elevates the potential of the group and, in doing so, the potential of the individual. But as the group expands, the risk or cost of failure is distributed more widely, and collective failure—the tragedy of the commons—becomes a more likely outcome. So size matters. In tackling collective problems, whether the climate emergency or our response to the next epidemic, we will be better served by relatively small and diverse groups focusing on region-specific issues than by national or global strategies that risk failure through intended or accidental exclusion of individuals.

Whatever our group behaviors have been recently, linguistic data suggest that something in our social outlook has taken a turn for the worse in the last few decades. A doctor diagnosing depression and anxiety in a patient might identify cognitive distortions, characteristic patterns of internalized thinking that manifest in overly generalized and typically negative interpretations of that patient's external world. Recently, researchers mapped out the changing prevalence over the last two hundred years of short word sequences reflective of these kinds of cognitive distortions. To do this they analyzed text from fourteen million books in English, Spanish, and German. Despite clear peaks of societal anxiety coinciding with major wars or periods of social unrest, the general trend in the use of negative language was, until 1978, "distinctly downward."[25] Then something changed. In the last four decades, the occurrence of word groups associated with distorted and negative thinking has risen at an accelerating rate. According to the data, we are now more anxious or more depressed (or both), than we were during either of the two world wars. And although the authors never mention climate nor make reference to worsening environmental issues, their data suggest that we, as a society, are increasingly stressed by the worsening state of our collective situation.

Our societies are increasingly sick, and the prognosis for us as individuals is not good. Among a myriad of other negative health impacts, we are increasingly subjected to traffic-sourced noise levels that impair cognitive functioning in our children, increase stress levels in adults, and suppress our immune functions in ways that may be responsible for premature mortality through cardiovascular and metabolic diseases.[26] We sleep less than we should, with notably shorter sleep duration prevalent among disadvantaged children compared to those from more affluent families. Sleep affects our ability to concentrate, to learn, and to form memories, and in missing out on adequate sleep, we increase our risk of obesity, diabetes, physical injury, mental health disorders, and drug misuse.[27] And, as with so many of the social feedbacks we've seen so far, the growth and expansion of our societies and the elevated technological prowess that we have unrelentingly pursued now threatens our social cohesion in far-reaching and non-local ways. Because one of the least visible consequences of our increasingly sleep-deprived societies is that we, as individuals, as community members, and as citizens of our respective nations, are increasingly less willing to help others. Why? During the last sixty years, the amount we give to charity relative to our income has steadily declined. This may be a response to changing social norms and economic structures, but it now seems likely that there is an additional contributor to the decline of the "helping economy." Helping others consistently engages a set of brain regions collectively referred to as the social cognition network. When we consider the mental states of others, their needs and differing viewpoints, we activate this network (comprised of the medial prefrontal cortex, mid- and superior temporal sulcus, the temporal-parietal junction, and the precuneus).[28] Functional magnetic resonance imaging, a technique that noninvasively maps brain activity based on changes in blood flow, reveals that even one night of sleep loss is sufficient to deactivate key nodes of the social cognition network. Compound that impact across the millions of people globally who regularly get inadequate rest, and the outlook for social harmony can seem worryingly bleak.

Joining the Dots

Thankfully, the competing forces that sometimes jostle and jolt our societies also impart energy for innovation and growth. Weathering the storm, therefore, just requires us to see what needs to be done and to find the collective wisdom to do it. In chapter 4, we saw that although bipedalism evolved in early hominids, it only became the modus operandi for ancestral and then modern humans because it enabled us to exploit new sources of energy for our rapidly growing brains. But as we saw, our brains didn't grow bigger because of the metabolic fuel we fed them; they enlarged in response to the pressure exerted by the increasingly complex societies in which we began to live. As the demands of social interaction increased, our cortex became more densely populated with the neurons necessary for sophisticated behavior. Obtaining the extra energy needed for this cognitive transition required group cooperation, and so a feedback loop became established that rapidly accelerated our physiological, intellectual, social, and cultural evolution. To investigate how this feedback might now help us cope with the challenges we collectively face today, we must unify disparate threads of an as-yet-unwoven cloth and explore the filamental connections of the fibrous social cloak that gives warmth to our stressed and anxious lives.

Social living has its challenges. Surrounded by other individuals, we must establish which of those others are likely to compete with us and which will cooperate. They move, as we do, through space and through time, living, loving, learning—adapting to change and responding to threats, optimizing their behaviors for the world that is theirs. How we keep track of this swirling cloud of antagonistic associations and fluctuating friendships depends on how and in which regions we more densely wire our brains. Studies on monkeys show that complex social environments drive growth of the neocortex, but growth of specific brain regions is not so much determined by how big the social group is; it is more specifically related to the number of meaningful social connections an individual has within that community. Social partnering preferentially develops regions of the brain associated with social decision making (such as the mid-superior temporal sulcus) and empathy (the ventral-dysgranular insula).[29] In essence, the more connected we are to each other, the better we become at managing those connections.

But just managing our connections as individuals would not have built Maeshowe nor would it have put Michael Collins two hundred thousand miles from Earth, all alone on the dark side of the moon. These feats of technological achievement were made real only because of something that went beyond the capabilities of a single mind. These giant leaps forward started life as fictional realities conceived, not by a single mind, but by a collective intelligence, an emergent phenomenon that first conceptualized the unthinkable then sang it into existence. Collective intelligence reflects the ability of a group to work together at a task—a statistical measure just like the intelligence quotient (IQ) applied to individuals. But the collective brain is much more than the sum of its members' individual skills, and because of this, a group's collective ability is not always easy to predict. We saw earlier—and revisit again throughout this and subsequent chapters—how in many dynamical systems (those multicomponent systems whose properties evolve through time as each component interacts with the others), it is the nature of the connections *between* the components, rather than the nature of the components themselves, that dictates their direction of change. And just as we saw earlier with examples of motoric synchronization, the level of congruence (in this case comparable skill level) between members of a group goes a long way toward explaining overall collective intelligence.[30] Groups whose members are in sync with one another, who adopt a clear group strategy, and who each contribute maximally to the task in hand are typically more successful than groups whose members adopt diverse approaches to problem solving and whose skill sets fail to complement one another. A key aspect here is that of skill congruence: members need not be experts in the same field, but there must be a point of overlap, a place where they might connect. When faced with a challenge, the congruent group must next explore the landscape of possibilities quickly if it is to be successful. When members actively engage in individual research that is then shared with the wider group, a rapid evolution to an optimal solution is likely. This is especially true when task complexity increases—groups of interacting people explore possible solutions more rapidly and generate far more potential ideas than those achieved by even the best-performing individuals working alone.[31] But when members instead rely on learning only from one another rather than from external sources, the chances of success are slim. Furthermore, this kind of "social learning" can precipitate entrenched views that become stubbornly protected within the group, even in the face of conflicting evidence and demonstrably more accurate interpretations.[32] In real life, social groupings often include a range of learning preferences—some individual and some social learners—so the evolution of the group's collective intelligence depends on the balance of the connections among group members. A feedback arises in which individual opinions shape the collective mindset, and the collective view in turn affects the opinions of the members. In a collective of individual learners, a few committed individuals with strongly held but unsupported views might struggle to make their views heard. But in a group in which learning from one another is more common, the embedded positive feedbacks mean that the vocal minority rapidly comes to dominate. In this instance, there is then only "one small step" to translate collective intelligence into collective behavior, which, through increased visibility, garners further support from social learners outside the group. Just as social media employs algorithmic feedbacks that preferentially promote popular ideas, people, or products, increasing their popularity and justifying further promotion, the positive feedbacks at work in collective behavior can, in the worst instances, rapidly escalate inequality and divisiveness even from foundations poorly supported by evidence.[33]

Although our Neolithic brethren most likely didn't consider themselves as nodes in a complex social network or components of a dynamical social system, their society evolved according to the same rules, the same feedbacks, that shape the modern social fabric we analyze today. Without a formal legal system or even formalized marital arrangements that, as we see later, only arose much more recently, subsistence societies of the prehistorical period would have had to self-regulate, meaning that each individual would have needed to consider not just how they could benefit from their community, but also how the community needed to benefit from them. This notion that the connections we make within a community have value both to ourselves and to others is an idea that we encountered earlier with the concept of social capital. Our web of friendships, familial links, coworkers, and so on determines our potential agency within the community to which we belong. But because of the interconnectedness of this web, we too contribute to the social capital of our connections—we are valuable to them, just as they represent value to us. And this value, in whatever terms it might be quantified, tangibly shapes our social prospects. Using the world's largest social connections dataset, researchers analyzed twenty-one billion Facebook connections to show that being childhood friends with people of higher socioeconomic status than yourself typically leads to greater economic mobility later in life.[34] The importance of social capital for escaping a cycle of poverty appears to arise from the interwoven and mutualistically interacting benefits of greater access to information, higher social aspiration, and exposure to job opportunities. Furthermore, when our social networks are viewed as multiple domains, or layers, of interacting networks—family, friends, online contacts, and professional acquaintances—it becomes apparent that even small modifications to connectivity in one domain can preferentially assist with connectivity in another. Coupling across our networks, as well as within them, harnesses connectivity feedbacks that not only enhance our social capital, but also promote cooperation, prosociality, and altruism.[35] Staying up late may be making us more miserly, but if we use that time to expand our social connectivity, maybe, just maybe, we can offset some of the damage done to our prosocial selves.

When "Could" Becomes "Should"

Sigmund Freud, the famous Austrian neurologist and founder of the field of psychoanalysis, held that civilization was born of two things: love and necessity.[36] Love enabled us to retain the closeness of our partner and the children that we raised together. Familial bonds and their shared realities, the values held in common between us, gave strength to the union and from this strength, repeated over and over in other families, came growth of a community. Yet external hardships—the need to provide food, shelter, and protection—required work, and the necessity for us to labor encouraged us to work together, to cooperate, to share a collective burden in a way that was mutually beneficial. Two forces thus arose. On one hand, we wanted to be free, to live according to our own rules, and with the people to whom we were devoted. But on the other hand, we knew we needed to play a role in society in order to ensure continued benefit from the collective strength of the community. Balancing these opposing ideas—I *can* do as I like, but I *should* do what is best—sets up a self-regulating system in which "could do" enhancing feedbacks are kept in check by "should do" limiting ones. Just as Zeus arbitrated among the feuding gods to resolve conflict and ensure harmony in the heavens, so too do we, as social animals, establish moral codes to guide our behavior. Through connections we accumulate social capital, but how this capital is "spent" is down to us. If our species were driven purely by a Darwinian urge to outcompete others, we would invest our social capital in ways that benefit ourselves above all others. Yet we are a prosocial and altruistic species; we are socially mindful to varying degrees, investing our social capital in relieving a burden from others or in improving a collective asset. Interpersonal benevolence, or charity, varies from country to country and appears strongest in countries whose performance on environmental protection is typically highest.[37] Individual actions raise country-level standards. And as we age we tend to engage more in helping behaviors, in part because emotionally stimulated older brains release more oxytocin, making us more trusting and generous.[38]

Altruism is, therefore, innate. And as long as we think of our actions as choices, rather than obligations, we feel stronger and more empowered. But from this empowerment comes a bolstered sense of self; we see ourselves more as independent and free-thinking individuals,[39] and such traits make our behaviors harder for others to predict. This means that the more we exert our right to behave the way we want rather than the way we know we should, the more energy is needed by those around us to connect and meaningfully engage with us. "Anticipatory anxiety" develops as those around us begin to feel the increasing uncertainty of not knowing how we might act or react at any given moment. More conformist individuals, by contrast, allow for greater prediction and more efficient information processing, which facilitates long-term planning and greater social stability.[40] This less challenging social situation ensures greater social cohesion, but from what we have seen in other chapters, it is also clear that the apparent dissonance between free thinkers and conformists may in fact be essential for effective social dynamics. By continually pitting our tendencies for individualism against our aspiration to altruism, the social dynamic that emerges has repeated opportunity to explore a suite of available possibilities, each iteration "learning" from the last and adapting its configuration until a solution is found that optimizes the outcome for both parties.[41] Our communal behavior therefore self-organizes to an optimal state. The same opposing pressures arise at the individual level too. As individuals we increase our network connections (friendships) over time, but it turns out that there is a maximum number of stable social relationships that an individual can maintain. This number, the Dunbar number,[42] is about 150 people. Originally derived from archaeological, evolutionary, and neurophysiological evidence, this upper limit on meaningful social connections has since been confirmed by complex social network models.[43] In these models, the benefits of larger social networks are countered by the increasing cognitive cost of maintaining those connections. Over time, size and cost compete with one another until a self-organized state arises in which information transfer within the network is optimized. The self-regulating feedback processes leading to social and individual self-organization in these examples are precisely the same processes we have seen previously, such as with the fluctuations of ice age climates or in the evolution of early language. In all of these contexts, the dynamically evolving system oscillates and iterates until it eventually settles on a "middle way" that indulges in neither extreme but is acceptable for all.

How then, in our self-organizing social networks, do new ideas, innovations, trends, and social movements gain traction and propagate? What role do social feedbacks play in social cohesion at one extreme and polarization at the other? Changing conventions and shifting norms arise when new ideas diffuse through our social landscape, but the rate and trajectory of that diffusion is controlled by the roles we each play in adopting the new alternative, stubbornly resisting the idea, or—for most of us—sitting on the fence until we see which way everyone else is going. When staunch advocates of a new idea attract other early adopters, uptake of that idea grows slowly at first. But eventually, a critical mass is reached that sparks rapid diffusion to the remainder of the population.[44] Why are we so easily swayed once this critical point is reached, when sitting on the fence was, to start with, perfectly fine? It comes back to the same kind of dissonance we saw just now, between free thinkers and conformists. So long as the majority of our social network is also sitting on the fence, we feel secure, because we think we can predict the behavior of the majority. But as the tide begins to turn and our shared reality becomes increasingly abandoned by those to whom we are connected, the network dissonance begins to increase, and our ability to predict the behavior of others starts to wane. Depending on how conformist we are, we might abruptly jump ship to the new shared values, or we might try to hold out just a little longer, but either way, we tend to seek the easiest way to reduce the dissonance we perceive.[45] Changing someone's mind, it seems, is often simply a case of giving them an easy "out."

The rapid diffusion of ideas when that critical mass of influencers is reached tips the system into a new and altered state. Identifying when and how such tipping points might be reached in social phenomena remains a challenge, but understanding what impedes such change—perhaps for beneficial purposes such as widespread action on climate and environmental issues—has a very real urgency. Changing social norms, as we have seen just now, relies not only on building a new set of shared values, but also in developing a common understanding of the benefits to be gained by adopting that new standpoint.[46] But what about the reverse? What if, instead of greater unity, increasing social dissonance triggers a regime shift that divides and polarizes us? The evidence suggests that political polarization is not so much driven by exposure to extreme opinions as it is by "unfollowing" those in our network that we sense have diverging views from our own. This gradual process of losing connection to part of our network insidiously separates us into two (or perhaps more) distinct groups whose members remain connected within that group but not beyond. From what we have already seen in this chapter, it is clear that, for prosperous societies, connection is everything. It is the source of our collective intelligence, our societal resilience, and our emotional and physical well-being. Understanding the feedbacks that control our connectivity, therefore, and learning to recognize when that connectivity is at risk must remain paramount in our social memory if we want to prevent tragedies of the past becoming disasters of our future.

Learning to Love

In this chapter we have progressively narrowed our vision from the great migrations of our ancestors, their early settlements and shared realities, to the birth of society and the unpredictable interactions of neighboring—sometimes conflicting—groups. Our social dynamical system emerged from a human substrate with an energy and a direction fueled and directed by both the individual and by the group. Locked in a spiraling evolutionary dance, we are like water in a social whirlpool that continues to grow, its form and fervor never constant, its position unfixed. But as this collective of individuals ebbs and flows, foaming in unison as our numbers rise and then collapse, we must now ask what motivating force it is that draws us to one another just strongly enough that we might find the elusive middle way and balance personal aspiration against cooperative union. What is it that makes us like each other?

Throughout recorded history, physical features like clear skin, bright eyes, and defined cheekbones have been considered "attractive," irrespective of age and culture.[47] But what use is beauty? Its appreciation becomes "mildly intoxicating," notes Freud, but it exists without obvious utility. And yet, as Freud goes on to say, "civilization would be unthinkable without it."[48] Beauty, like love and necessity, has enabled us as individuals to see something beyond what is immediately apparent, to sense the divine in the ordinary, order amid chaos. The artist Robert Henri describes beauty as "the sensation of pleasure on the mind of the seer."[49] The Greeks devoted enormous energy to the appreciation and understanding of what makes something beautiful and what value that has for shaping our lives.

Aristotle argued, for example, that happiness was the ultimate goal of existence, whereas Epicurus felt that the highest goal to which we could aspire was pleasure, in the sense of freedom from pain and fear.[50] But what if, beneath the veil of philosophic musing, there were a physical, observable basis for our attraction to what we find "mildly intoxicating"? And what if that observable thing exercised some agency in a loop of adaptation and refinement, a feedback that propelled us incrementally forward toward a better, more optimal state?

Infectious diseases have been the greatest cause of human mortality throughout history, a selective pressure that should plausibly have favored survival of genes whose hosts were more successful at avoiding such an end. Sickness can manifest in many ways, but we can separate acute responses (such as localized inflammation) from chronic, or systemic, responses that arise from long-term suppression of the immune system. As it turns out, there is no particular correlation between perceived facial attractiveness and measures of acute inflammation—we don't seem to mind if our mate has a short-term health problem. But there *is* a link between attractiveness and blood-based indicators of antibacterial immunity—factors that reflect much longer term bodily health. So in being drawn to those we perceive as more attractive, we are, perhaps without realizing it, subconsciously but deliberately choosing healthier partners with more highly functioning immune systems.[51]

From an evolutionary and even social perspective therefore, we unconsciously use beauty as a way to parse what is accessible to us and select from that set the partners who are most likely to survive longest and provide our genes with the best chance of longevity. Through beauty, we seek immortality. Unsurprisingly, this conclusion, based on modern quantitative science, echoes the wisdom of Plato, who, nearly two-and-a-half thousand years ago, said more or less the same thing: "the object of love is not beauty. . . . It is birth and procreation in a beautiful medium. . . . Because procreation is as close as a mortal can get to being immortal and undying."[52] But is social pair-bonding really all just about future generations, about maintenance and proliferation of our genetic code? If that were true, it would be a far safer bet to reproduce asexually. By bringing another body into the equation, we dilute our genetic material and make ourselves vulnerable to physical harm or infection from our partner. Yet exclusively asexual reproduction occurs in only one in a thousand animal species, so clearly there's a benefit to sex. Traditional wisdom asserts that it is only by combining our genetic material with others that a species can adapt and evolve. But what if there were also a benefit to us, while living, and not just to our descendants? In algae, environmental stress induces sex, and in activating the processes necessary for sexual reproduction, the algae become better able to deal with whatever kind of a stress it was that triggered the act. In other species, such as fruit flies, it appears that mated females exhibit greater resilience than their virgin sisters, through stronger immune responses.[53] Prostaglandin, a hormonelike compound common in many animals and transmitted through semen to sexual partners, may be the source of this immune boost, and by also helping with egg development, can increase reproductive success.[54] Perhaps even more remarkably, there is evidence to suggest that mating in rodents can both boost cognitive abilities in the short term and slow cognitive decline in the long term.[55]

Viewed through the lens of the beneficial feedbacks we appear to glean from reproductive pair-bonding, it makes sense that we would adopt a social lifestyle that facilitated our access to such benefits. Anthropological evidence suggests that, prior to about four thousand years ago, family groups consisted of a loose arrangement of a couple of dozen people, in which there were several male leaders, multiple women, and numerous children. In light of what we now know about reproductive benefits, this kind of arrangement makes perfect evolutionary sense. But as Neolithic communities swelled and encroached upon one another's territory, the concept of land ownership applied pressure to this congenial state of affairs. Continued survival of our genes then began to rely on something more than simply passing them on—we had to provide security for the bearers of our genetic material. Marriage became widespread among the ancient Hebrews, Greeks, and Romans, but it was hundreds of years later before it became religiously aligned, and hundreds more years before romantic desires, rather than property, became the motivating force for long-term pair-bonding.

We are, however, still left with a problem. Beauty may help us select the most beneficial mate for our aspirations to immortality, but to maximize the benefit of that mate, both through reliable access to the immune-enhancing reproductive act, as well as the repeated opportunity for genetic dispersal via multiple children, we need to ensure that our connection to that mate is maintained. If there were a unique secret to a long and happy marriage, someone, somewhere, would most likely be very rich. Unfortunately, it seems that such a singularity flies in the face of what we anecdotally "know" to be true, that the basis of long-term attraction and marital satisfaction is different for every partnership. Some swear that "opposites attract," whereas others prefer "birds of a feather." Our histories, lifestyles, personalities, socioeconomic status, and many other factors besides, all contribute to what makes us *us*, and, given the infinitude of possibilities, combining any two people's myriad characteristics makes it highly unlikely that we will ever meet a partner who is either entirely similar to—or completely dissimilar from—ourselves.

So if our marital synergy isn't so much about our traits, what is it that makes some couples tick, while others fall apart? When we watch a film or show together, our brains process the same information stream as our partner next to us. That stream comprises auditory, visual, and emotional signals as well as a host of other diverse stimuli. But in processing this shared narrative, our cerebral responses sometimes fall in step with each other, allowing neural synchronization to develop right across the brain. And when those synchronized behaviors are analyzed in the context of marital happiness, the evidence indicates that the strength of coordinated real-time responses to interpersonal or socially relevant cues is a good predictor of relationship satisfaction.[56] Furthermore, stronger synchronization in a connected set of brain regions known as the default mode network (DMN) correlates with how similarly a couple might deal with financial management, how they communicate, and how egalitarian they are about their respective roles. Put simply, marital bliss isn't so much about being on the same page as one another; it's about wanting to turn the page at the same time and for the same reasons.

Timing, if not quite everything, is therefore surely important. In controlled experiments, researchers measured social connectedness by tracking the time taken for communicating individuals to respond to one another. Feeling connected is critical for our emotional and physical health, so anything that mediates interpersonal connection can affect our well-being. When conversations feel fluid, when we feel we're really engaging with another person, our response times can be extremely short, around a quarter of a second or so, far quicker than could be achieved through conscious control. So our brain, to some extent, operates on autopilot. The rapid responses we give are built from a complex suite of predictions, ones that not only attempt to prepare a suitable response sufficiently far in advance, but predictions that also try and anticipate when our turn is coming, when and how we should deliver our reply, and what the reaction of our partner might be.[57] As the conversation progresses, we try and keep track of the dialogue in order to see where it may be headed and what our partner might say next. The latency in our conversational exchanges therefore reflects the degree to which we can read each other, which itself is a measure of how strong our connection is. Intriguingly, however, we hold ourselves to different standards than those we communicate with—we infer stronger connections based on how quickly our partner responds to us, rather than on how quickly we respond to them. Is it okay to leave them hanging? Perhaps a few extra milliseconds doesn't matter too much, but in the age of increasingly online communication, the lag between communications can stretch into days, weeks, or months. Or perhaps they disappear completely, like a ghost. As a relatively new phenomenon, social "ghosting" has yet to be fully researched. But the evidence that does exist suggests that being ghosted by a friend can, in the short-term, leave a person with feelings of overwhelming rejection, confusion, and wounded self-esteem, and in the long term breed mistrust, self-blame, and feelings of worthlessness.[58]

Looking at the Sky

The transactional language we've engaged in above is the language of quantitative research, of correlations and tendencies, of dependent and independent variables. But these are not things that are real to us, in the sense of being aware of them. In learning how to interpret the world as an artist, Robert Henri entreats us to simply, "Paint what you feel. Paint what you see. Paint what is real to you."[59] In the next chapter we explore in more detail what it means to be us, to establish what is "real." But there is something very real that has shaped every member of every society from birth, a connection that defines from our very first moments what we feel and what we see. And that is the connection to a mother.

Feelings of social connectedness and bonding, as we saw earlier, arise in specific regions of the brain. The prefrontal amygdala networks bring together regions responsible for higher order learning and decision making with areas that govern emotion and valuation. Often when we "click" with someone, we feel a surge in connection when our eyes meet, when we gaze rather than look, when peripheral distractions dissolve. "The mind," said philosopher Alan Watts, is like "an eye that sees, but cannot see itself."[60] Locked in a mutual gaze with another, our brains engage in distinct patterns of behavior, neurons firing across multiple regions of the brain depending on the context and evolution of the gaze. From the "social gaze," we derive meaning and information—the eye and the brain each "see" in their own way. At birth we can't focus our eyes sufficiently well to make such connections, but once those first exchanges occur, the powerful combination of shared gaze and deliberate vocalizations synchronizes mother-infant brain activity and fosters the development and embedding of complex social concepts such as collaboration, emotional regulation, and empathy.[61] Early life neural coupling like this sets the stage for longer term development and learning, in particular the maturation of the right hemisphere of the brain. The right hemisphere controls functions essential for our survival, such as our ability to identify and respond to stimuli in the space around us, interpret social information, and regulate emotions. Until age three, the right hemisphere dominates. Through interbrain synchronization using right hemisphere theta rhythms, mothers tune an infant's social brain, preparing it for the challenges of life ahead.

From our birth, we are like spring water that pools then starts to flow. The babbling brook of our formative years imparts to us life and exuberance, a vitality of unknown possibilities and unclear intent that swells and matures as we enrich the stream of our life with the connections of social ties. Our path is ever onward, driven by a gravity outside of ourselves and yet mediated by a steepening or shallowing bedrock of happenstance that accelerates or restrains us, by a downpour that engorges us more quickly and uncomfortably that we might wish, or by periods when we long for the rains to renew and revive us from the withered, weak, and lifeless runnel we've become. Those droughts are the difficult times when we can carry no great load, can afford no support to the ark of the collective that might turn to us expectantly. But replenished and in flood, we burst along excitedly, full of promise and fertile with ideas. Part of a swollen whole, we have such momentum that even boulders protruding the riverbed only serve to lift us up, above the surface, to gulp greedily at the air and to sink back below, refreshed and enlivened. We are not the same river; we are not the same (hu)man.

But for how long, we might ask, might this mighty meander persist? Our immortality is in our children and in those that come from them; it is not in us. Because even if we survive the onslaught of disease, injury, or accidental death, our bodies keep score of the life we have lived, the scars we have won from the battles we have fought. Mortality and chronic disease increase exponentially, doubling every eight years.[62] From age forty, we begin to deteriorate ever more quickly, such that the slow and steady aging that reduces our bodily resilience eventually hits bottom and our aged river is finally subsumed by the ocean from whence its waters first came. According to a recent study based on blood markers,[63] there is an upper limit on how long a human body could theoretically survive. And that study suggests that we can expect no more than 120 to 150 years on this planet, the absolute limit on human life span. But though we might not be able to make those years longer, we can perhaps make them better. The remarkable properties of our gut microbiome mean that we carry with us the potential to recover more quickly from antibiotic disturbance, boost our immune system, or even rejuvenate the aging brain[64] by "stool banking"—an emerging service that preserves the healthy microbiome of your fecal matter for later intestinal transplant when needed. If that sounds too unsavory, then perhaps transcranial alternating current stimulation might be preferable? By applying specific frequencies of electrical currents through the skull to specific regions of the brain, rhythmic theta- and gamma-frequency brain activity can be manipulated to boost both working and short-term memory.[65] Easier still, we could eat less. Multiple studies show that limiting the number of calories we consume reduces or delays diabetes, cancer, and heart disease, while also promoting the maintenance of both lean muscle mass and brain volume.[66] Intermittent fasting might be even more powerful. Even when the total calories consumed are the same, periodic fasting is superior to other forms of caloric control when it comes to enhancing memory, most likely because it triggers new brain cell growth in the adult hippocampus.[67] Thinking back to chapter 4 and the transition from continual grazing to more episodic ingestion of calorie-dense meat, we might wonder whether this change in feeding frequency, rather than the source of the calories themselves, might have been instrumental in *Homo*'s accelerated increase in brain size. Either way, a feeding pattern that follows our circadian clock, that is, eating only when we are most active, has recently been shown to afford many physiological benefits, including weight loss,[68] and collectively the benefits of these dietary hacks seem to lie at least partly in an increase to our physiological resilience by imposing a mild level of cellular stress that triggers our bodies to adapt and improve—a theme that is, by now, no doubt familiar. By choosing how and when we eat, we control the direction of the adaptive feedbacks that can either send us to a premature grave or ensure a longer, more cognitively and physically capable life.

In our earliest societies we learned to connect with, empathize with, and support one another. We used language to build culture, used culture to share ideas, art, and science, and we united with one another for strength and for prosperity. We harnessed the feedbacks that helped us to grow by devising mechanisms that enabled us to better predict natural cycles, group dynamics, and personal behavior. But despite our progress, we still look up at the stars in wonder and feel the smallness of our individual self and the fragility of our existence. Our social organism is a complex, dynamical system composed of eight billion moving parts, but for how much longer? Fertility rates in the United States, China, Germany, Italy, Spain, and Japan are well below replacement, meaning that our global population is likely to peak in the next few decades and decline by the end of the century.[69] Political and economic instability tend to rise and fall with irregular oscillations because of the complex feedbacks embedded in these social systems. But we are under pressure, now, in ways that have seen rising social anxiety and discontent. Homicides in US cities jumped 44 percent between 2019 and 2021, and half of Americans recently surveyed expect a civil war "soon." Seven percent of those surveyed said they would be willing to kill other people to advance what they considered to be an important political objective.[70] And social divisiveness isn't just a stateside phenomena. Environmental protection measures have enabled the remarkable recovery of wolf populations in Central Europe over recent years, yet local residents now perceive these animals as a threat to their livelihood, and in areas where wolf attacks have increased, communities now vote for far-right parties.[71] Knowing what we know about the importance of social connectivity, of polarization dynamics and tipping points, of unfollowing, ghosting, and othering, knowing what we know, it is time we look more deeply inward, to get to know who we really are, to get to know our *Self.*

SELF   6

When I let go of what I am, I become what I might be.

—LAO TZU

Right now, your body is aglow. You are illuminated from within by biochemical reactions that emit photons, massless elementary particles that carry the electromagnetic force at three hundred million meters per second in a wave-particle unity. We are radiant, illuminating beings, shining a light that rises and falls smoothly throughout the day, peaking in the late afternoon. More delicate than the bioluminescence that makes glowworms glow or phosphorescent algae shimmer, our light comes from metabolic processes that trigger fluorophores—fluorescent chemical compounds—to release biophotons. The light that is given off is a thousand times too dim to be visible to the naked eye, but it exists and can be photographed. Using a camera cooled to 120 degrees Celsius below zero (−184 degrees Fahrenheit) and sensitive enough to capture a single photon, researchers took images of human subjects throughout a day in the visible parts of the electromagnetic spectrum.[1] For comparison they simultaneously imaged these subjects with a thermally sensitive camera, capturing only the infrared components of the electromagnetic spectrum. Comparing the visible and the thermal images, they saw that although heat was primarily radiated from our necks, it was from our faces that we shone and shared our light most brightly. If only our eyes were sufficiently sensitive, we would see things that we normally do not perceive; we would see a body that "glitters to the rhythm of the circadian clock," a single beacon of light among billions of other pulsing, energetic, radiant beings.

In Mahayana Buddhism, the concept of light and luminosity appears in the "state of no-mind," the ultimate reality in which we are free of delusions and exist in a state of complete awareness. This mind of clear light is a luminous mind, awakened to and understanding of the radiance of pure reality. The *Bardo Thödol*, the Tibetan Book of the Dead, describes the encounter with this state in the process of dying: "Thy breathing is about to cease. Thy guru hath set thee face to face before with the Clear Light; and now thou art about to experience it in its Reality in the *Bardo* state, wherein all things are like the void and cloudless sky, and the naked, spotless intellect is like unto a transparent vacuum without circumference or center. At this moment, know thou thyself; and abide in that state."[2] And it is in this moment, between the last outbreath and the last inspiration, when the invisible force of life remains in the "median nerve," that the state of knowing will be achieved. *See you on the other side.*

Light, therefore, is a wonderfully metaphorical phenomenon—on one hand a physical manifestation of wave-particle electromagnetic energy, and on the other, a metaphysical concept associated with awareness, reality, and transcendent understanding. And so too is it when we come to think of our "Self." For the inseparable unity of our *self* can be considered both in relation to our physical body, our oneness and individuality in terms of the matter of which we are composed, and also as our "sense of self," the thoughts, emotions, and beliefs that make us who we think we are. That we can be aware of either of these things separates us from the majority of other organisms. Many animals are sentient (able to respond emotionally to sensory inputs), from crabs, cuttlefish, and octopus right up to higher mammals such as humans. But only a few creatures, such as dolphins, elephants, magpies, and some apes, recognize their own bodies—they have a sense of self in physical terms. Do they also perceive their inner self, their invisible self?

Although we cannot yet answer such a question, it is the separation of phenomena that can be physically experienced from those that can only be felt or understood in a nonphysical way that underpins our inseparable bipartite self. In the words of E. F. Schumacher, "experience, and not illumination, tells us about the existence, appearance, and changes of sensible things . . . while illumination, and not experience, tells us what such things mean."[3] Recognizing and reconciling the disparate needs and desires of our dichotomous single self is the journey upon which we each embark at birth and the challenge with which we live until death releases us into the "void and cloudless sky."[4] To know the world around us, to know our physical self in the physical world, is the job of our senses and of our organs. But it is quite different from knowing our nonphysical self in the nonphysical world, which relies on self-awareness, a deeper understanding, and a theory of mind that allows us to contemplate what others might be thinking, and what our influence upon them might be. Yet so intertwined are these two, and so important is the nonphysical self, that some form of this concept is common to nearly all ancient and historical traditions, from the medieval European mysticism of the *Theologica Germanica*, to the first millennium BCE *Tao Te Ching* of the Chinese philosopher Lao Tzu. And of course, "Know thyself" was also the first of the three Delphic maxims inscribed at the entrance of the Temple of Apollo at Delphi, the site where ancient Greeks beseeched the gods for prophesies and protection—an indication that, to those who had not yet sought inner wisdom, even divine knowledge would be meaningless.

Clocks, Nerves, and Pacemaker Cells

The wave-particle nature of quantum phenomena is widely acknowledged today, but the idea that light moved through space as a wave can be traced as far back as the seventeenth century to Dutch mathematician and astronomer Christiaan Huygens. He first wrote of the idea in 1678 and eventually published it in his *Treatise on Light* in 1690. Huygens's idea was rejected by Isaac Newton, however, who instead preferred to interpret light as being made up of discrete particles. Reflected light, according to Newton, behaved similarly to a ball bouncing against a surface—the angle at which the ball (or light particle) left the surface was the same as the angle with which it first collided with that surface. Two centuries later, Albert Einstein built on the electromagnetic insights of James Clerk Maxwell and Max Planck and showed clearly that light indeed behaves as a wave, just as Huygens had proposed, but also that light can equally behave as a stream of particles, as Newton supposed. The inseparable twin forms of nature, it turned out, was something we needed to learn to live with.

But some years before his work on light, Huygens made another, quite different contribution. He noticed that two pendulum clocks, when resting on a floor supported by a common beam, tended to synchronize with one another over time. The "sympathy of two clocks," as he called it, has now become known as "Huygens synchronization" and seems to occur in a wide range of natural dynamical systems.[5] What is necessary for this phenomenon to take place is the subtle but continued feedback between each of the ticking clocks or, in other systems, between two or more coupled oscillators. The forces exerted by each oscillating component are felt by the other components, and in a tiny way, those other components are disturbed by the transmitted force. These disturbances minutely change the swing of the pendulum, or the frequency of the coupled oscillator, to the extent that over time these changes "entrain" the weaker component with whichever one is beating more strongly, until there is no longer any difference between them.

Our brain consists of tens of billions of neurons, and groups of neurons tend to behave as oscillators, each group rhythmically fluctuating in activity at frequencies that differ from one group to another. Yet given the correct impulse, these connected but independent groups of neurons can be entrained and synchronized, just like Huygens's clocks. Through the synchronizing effects of electrical feedback, neurons can align the frequency of their activity in ways that allow them to exchange information more efficiently, enabling critical functionality of our physical self, such as learning, memory formation and retrieval, or simply just perceiving and paying attention to the physical world around us. And it is this "pacemaker" mechanism that gives rise to the so-called theta rhythm of our brain from a region known as the hippocampus, named after the Greek mythological "seahorse" to which its appearance was thought similar. The purpose of the hippocampal theta rhythm is not fully understood, but it has been linked to thinking and moving activity, such as memory and navigation. Bat brains exhibit bursts of theta wave activity during echolocation, and in mammals such as humans, bursts of theta wave activity seem to occur (both from the hippocampus and the brain's outer cortex) during the rapid eye movement (REM) phase of sleep and upon waking. Theta wave activity also seems to be enhanced through meditation, with some practitioners thus being able to reach the same dreamlike state of waking from sleep while still fully conscious.

Synchronized neural connections therefore underpin key parts of the physical manifestation of our self in ways that also naturally impinge on the nonphysical aspect of our self. Without the feedbacks that allow neuronal synchronization, regions of the outer layer of the brain—the cerebral cortex—that control vision, touch, and movement wouldn't be able to communicate with each other effectively, and tasks such as catching a ball would be almost impossible. To ensure that these cortical components are wired correctly, the neurons of our brains mature at different rates, so that those responsible for touch are ready first, allowing a newborn infant to suckle, for example, whereas those required for movement and sight come later.[6] But how do these processes of neuronal synchronization compare with other feedbacks we've seen in previous chapters? Perhaps surprisingly, the mechanism that dictates the way our brain communicates within itself is very much the same as the slow buildup and rapid collapse we saw in the tectonic and rainfall patterns of the early Earth of chapter 1, in the sawtooth pattern of species overturning in chapter 2, and in the ice age climatic oscillations of the last million years that we encountered in chapter 3. In each of those examples, we saw how a system could be gently nudged by an external force, slowly but steadily, until it reached a critical point beyond which any further nudging triggered rapid and trans-formative change, returning the system to its original state and allowing the asymmetric cycle to start anew. And so it is also in the brain. Neurons (nerve cells) tend to communicate with one another by way of *action potentials*, short-lived pulses of electrochemical energy that arise from the steady buildup of ions of salts such as sodium or potassium. Impulses travel along axons—elongated parts of a nerve cell—as the accumulation of these ions sequentially open and close special channels in the cell membrane, driven by the difference in electrical charge between what is inside and what is outside the cell. Eventually the signal reaches the end point of the nerve, the synapse, where it triggers the release of another chemical, a neurotrans-mitter, which initiates the same process in the next neuron, and so on.

This "integrate-and-fire" mechanism is a way of absorbing a (rela-tively) slowly accumulating external energy source in a way that allows that energy to build up (integrate) to a level necessary to do some work (fire)—just like charging a battery. The advantage of such systems is that they allow for the progressive synchronization of multiple components, because even though each component might initially be charging from a different state, if they are close enough to the critical threshold when a neighboring component fires, they too will be triggered.[7] The combined energy released might then be sufficient to trigger others that are also close to the threshold, producing a cascade of firing that, over repeated cycles, gradually entrains each and every discrete oscillator and self-organizes the system as a whole. Just as Huygens had noted with his pendulum clocks three hundred years ago, modern neuroscience has revealed the stunning simplicity and elegance of the feedbacks governing the brains and nerve cells of our physical selves and the vital role they play in keeping us alive.

And what could be more emblematic of "staying alive" than the beating of our hearts, the rhythmical throb that unerringly propels blood through our veins and entreats every facet of our vitality to persist? From the Latin *pulsus*, the past participle of the verb *pellere*, meaning "to push," or "to drive," we have *pulse*, the manifestation in our physical selves of the invis-ible something that differentiates living from nonliving, that fundamental essence—life—that cannot be easily defined. It comes not simply from the cells of our bodies, yet it flows through them and invigorates them. But it also isn't in that flow of energy, for many systems are governed by a flow of energy and yet are not thought to be "alive"—the coursing of a mountain stream, for example, or the slow-motion degradation of a mountain hillside over eons. The inexpressible something that constitutes life is once more both a thing seen and yet simultaneously unseen, a single phenomenon that is ever familiar and yet still strange. Later in this chapter and in those that follow, we expand upon this glimmering light of existential uncertainty, but for now, let's return to ion channels and the electrochemical "light" that pulses and strokes our physical being.

The control room of our lighthouse, the physical side of our self, might be our brain, but the engine room that keeps us illuminated is the heart. Our heart is a muscular organ about the size of a closed fist, with four chambers that alternately swell and contract as they fill with and expel blood in a precisely coordinated sequence that typically lasts just less than a second. On average, we might expect our heart to beat like this two or three billion times during our life.[8] But if diseased, our engine of life loses its normal rhythm and degenerates into faster, irregular behavior that marks a breakdown in the carefully regulated electrical activity of the heart. Just like the neurons seen previously, effective functioning of the heart depends on the synchronized activity of a complex suite of components, coupled oscillators uniting nervous impulses with muscular contractions repeatedly switching on and off to meet the evolving demands of body and brain. And just as with the brain, the pacemaker cells of the heart rely on voltage-gated ion channels to integrate an accumulating electrical charge and then fire it from one nerve junction to the next in a cascade of near-instantaneous electrochemical signaling. The pacemaker cells reside in part of the upper right atrium of the heart, known as the sinoatrial node. Cells in the sinoatrial node are electrically coupled, meaning that as the charge state of each cell increases toward the same kind of firing threshold we saw before with nerve cells, they become mutually interdependent, entrained with one another and sufficiently synchronized that no dominant pacemaker cell is required, only the collective dynamics of the population as a whole.[9] Like the regime shifts we have seen earlier in the rapid explosion of life during the Cambrian geological period or in the social phenomena of chapter 5 that underpin collective responses to natural hazards and to changing societal norms, once the electrochemical state of the heart reaches a critical point, the release of its energy can be triggered by any single cell of that node. Pulsing a wave of contraction first through the walls of the atria and then through those of the ventricles below, our blood is pushed through sixty thousand miles of arteries, veins, and capillaries to deliver oxygen and nutrients to every cell in the body.

The heart system we know today, therefore, is one that is synchronized, self-organized, and needs to reach a critical state in order to function correctly. The charge–discharge cycle of the sinoatrial node cells, the pacemaker of this magical organ, reflects the gradual accumulation of electrical potential followed by its sudden release, an asymmetrical fluctuation familiar to us now as the relaxation oscillator we've seen in earlier chapters, a reflection of the fundamental similarity in the feedback processes that regulate systems as diverse as heart cells, species evolution, and climate change. But the heart has still another secret to share. In chapter 2 we learned of the Sheldon spectrum, the distribution of biological mass of organisms on Earth that reveals a property we referred to as scale invariance, a fractal pattern in which mathematical relationships look the same regardless of the scale at which they are measured. Here too in the heart, we can see the same natural order of things, a fractal geometry in the variability of a healthy heart rate that spans three orders of magnitude.[10]

But though the heart powers our light and the brain directs its beam, our beacon is more than just these two. The upright pillar of our bipedal frame also houses organs that take care of our maintenance, storing what needs to be stored, cleaning what needs to be cleaned, replenishing our bodies with sustenance from beyond our walls. To keep the lantern alight, a complex ensemble of interdependent components work endlessly in their united aim, but what form does their communication take, what messenger relays their missives? This monumental task of coordination and regulation relies on many things, including the gut microbiome that we explored in chapters 4 and 5, but the main superhighway that runs through our body is the vagus nerve, a meandering network composed of more than one hundred thousand fibers that couple almost all internal organs to our brain. The vagus nerve is part of our parasympathetic nervous system, the web of nerves that maintains our normal functioning, and as part of that web, the vagus nerve controls automatic behaviors such as breathing and digestion or the beating of our heart.

With its toes dipped in the waters of our vital organs, the vagus nerve relays messages to the base of the brain, where the aggregated signals are deciphered and interpreted to form a picture of the internal state of our body. This sensing of our inner physical self goes by the name of *interoception*, and it's a two-way street. Sensing the body's needs from the interoceptive signals it receives, the brain replies with adjustments to be made, perhaps anticipating ahead of time what an organ might soon need. The conversation between the brain and the organs therefore relies heavily on feedback, either positive feedback to rapidly increase functioning in a particular way or negative feedback to return an organ's behavior to a less active state. This process of autonomous regulation has close similarities with the concepts of "control theory," which arose in the mid-twentieth century in the field of electronics and electromechanical engineering. Promoted and popularized to a large extent by the work and publications of the American mathematician and philosopher Norbert Wiener,[11] who coined the term "cybernetics" to describe the systems he envisaged, the control theory of feedbacks in inanimate physical systems set the stage for James Lovelock's Gaia hypothesis a quarter of a century later. Lovelock put forward the idea that Gaia, the Earth and all its myriad systems, worked as a single entity to maintain a balance that favored proliferation of life.[12] Wiener and Lovelock differed in their areas of interest, but the mechanisms were the same. Individual components of a complex system provided feedbacks to the other elements of the system to which they were coupled. Control theory proposed that, under certain circumstances, these feedbacks could allow a system to self-regulate. That is, if one part of the system is tending toward an unsustainable state, another part might exert enough of a limiting effect to stabilize the errant component and ensure overall stability, just like a thermostat controlling the temperature of an oven or a house. Wiener went into some detail regarding the similarities between the human nervous system and "computational machines" and theorized that all intelligent behavior was the consequence of feedbacks.

An important requirement for Wiener's control systems was memory. A system relying on feedbacks to achieve self-regulation would be increasingly likely to fail if it operated simply by reacting to incoming signals without also benefiting from some insight as to what the previous state of the system had been and how it might have responded to past signals. Recalling E. F. Schumacher, we might remember that "experience" must also be matched with "illumination"—knowledge and understanding. Or, as Alan Watts notes, "every feedback system needs a margin of 'lag' or error" to avoid it forever switching on and off with such sensitivity that it is reduced to a trembling state, "startlingly like human anxiety."[13] This lag, or memory, is where learning takes place. Our coupled body-brain feedback system is a learning system, and the role played by the vagus nerve in this cognitive process has been examined experimentally. Researchers have shown that learning of motor tasks can be accelerated if the brain-body feedbacks are strengthened using a technique known as vagus nerve stimulation (VNS), in which the vagal nerves are directly stimulated with an electric current.[14] As the connector between the part of the body carrying out the motor activity and the brain receiving and replying to signals from that particular limb, the vagus nerve plays a key role in mediating the strength of the feedback. With VNS, it enables reinforced learning that consolidates the memory of the completed task. But there's more. Because the vagus nerve is part of a much larger interoceptive brain network that not only regulates normal body maintenance, but also controls memory, emotional processing, and the way that we formulate our own sense of self, it has been shown to play a key role in relieving stress, anxiety, and depression.[15] In infants, the vagus nerve plays a central role in communicating to the brain the body's experience of touch and being touched. Children "learn to be human through touch"[16] and rely on comforting positive touch experiences to feel safe and secure. But beyond the immediate response there lies a hidden feedback, one that shapes a child's development and behavioral disposition. Positive touch—being held affectionately, cuddled, and caressed—stimulates the vagus nerve, releases oxytocin, and decreases cortisol. In turn this reduces social anxiety and encourages openness in social interactions later in life. But negative touch, such as smacking, communicates rejection. And a rejected child becomes a less regulated and less cooperative one, a child who is less socially engaged and develops externalizing behaviors such as aggression and defiance. The body-brain connection, by way of the vagus nerve, is therefore critical not just in controlling our present physical self, but in also shaping our future nonphysical self. The adage "healthy body, healthy mind," attributed to the Roman poet Juvenal, seems, therefore, to have very real foundations. And thankfully, one of the best ways to self-stimulate the vagus nerve might be as simple as taking a deep breath, just as Buddhist meditation or the practice of yoga might teach us. *Long breath in, slowly out.*

Making and Shaping a Brain

Our brains start forming from about the third week of gestation, as neural progenitor cells divide and differentiate into the two different cell types that comprise the foundation of the nervous system: neurons (nerve cells that produce and conduct electrical signals) and glia (nonelectrical cells that play more of a structural or supporting role). By week ten, the fundamental components of the neural system have been established. We saw previously how different regions of the brain develop at different stages in order to maximize the connectivity of the areas whose tasks are needed earliest. And to start with, the infant brain more resembles a collection of fragmented modules than one that is widely connected.[17] But gradually, the multitude of task-specific brain regions are connected by an intricate web involving millions of nerve cells, each axon and synapse transmitting electrical signals to the next in line. Functional networks are built that connect disparate regions in ways that help them carry out multitask processes. First, it is the primary networks that are established, enabling basic functionality to support movement, visual, and auditory processing. Subsequently a suite of "higher order" networks develops that facilitates more advanced cognition, using memory to coordinate thoughts and actions. To be effective, functional connectivity relies on the time-coordinated activation or deactivation of separate brain regions so that all the necessary processing tools can be brought to bear on the particular task at hand at the same time. In the maturing infant brain, differences in connectivity within these networks appear to predict differences in childhood temperament and behavior.[18] In essence, the wiring that takes place in the first month of life seems to set the stage for how we behave, and it might even shape the kind of person we become. It is in these early stages that reinforcing feedbacks allow brain regions to rapidly connect and to stay connected. Neurons of the same age exhibit distinctive patterns of connectivity and activity, suggesting that those that are "born together, fire together."[19]

By measuring changes in blood flow through the brain, these networks of coordinated activity can be mapped out in our physical self and can be observed in ways that allow us to make inferences about those aspects of our nonphysical self that somehow emerge organically from the neuronal machinery. We know, for example, that the frontoparietal network (FPN) controls attention, and both the FPN and the homologous-interhemispheric network control or regulate emotions. Perhaps the most important network of all, however, is the default mode network (DMN). Later in this chapter we delve more deeply into the DMN, but for now it is sufficient to note that coordinated activity in this ensemble of brain regions is most prevalent when we're not engaged in complex tasks, but instead let our mind wander. Together, our suite of functional brain networks underpins the way we receive, process, and interpret sensory signals, and from that morass of data we derive information that keeps us alive. But just as a light that flickers or moves provides more information than a constant one, so too does the brain rely on both the constant functioning of its networks and on their flickering—switching on and off between modes of operation. The DMN, for example, is sometimes referred to as the "task-negative network," because it is most active when cognitive attention isn't being directed toward a specific task. As we see later, hyper-connectivity within the DMN becomes disabling, trapping an individual in a world of self-generated thought and isolating them from the tasks of everyday life that demand more focused attention. Switching, therefore, gives us flexibility and enables us to direct metabolic resources where they are most needed at any given time. Through switching, our brain optimizes its functionality and enables us to rapidly adapt to our ever-changing situational environment. And it's not just while we're awake that this switching occurs. Because every time we release our grip on consciousness and slip into sleep, our neural equipment engages in a competition for resources that pits working memory against long-term memory. Both are essential to our normal functioning but are served by antagonistic mechanisms. So when long-term memory is active, demanding access to our vast store of knowledge and experience, our working memory shuts down, and vice versa.[20] But rather than limiting our cognitive capacity, this on/off modality provides us with a fundamental benefit, because it is precisely during the "offline" periods that performance gains can be made—just like a computer installing an update late at night, when its user is asleep.

So far we've considered brain networks simply in terms of the large-scale functional networks that dominate much of our neural activity. But beneath this simplification lies a pattern that is, by now, familiar. Just as we saw previously with the scale-free pattern of heart-rate variability, patterns of brain interactions also display this kind of fractal relationship: there are a few large networks, a greater number of smaller ones, and many more even smaller ones, and so on. And despite the potential complexity arising from our one hundred trillion neuronal connections, it turns out that the networks that arise are similar from one person to the next, despite these smaller networks being hastily assembled, on the fly, as and when they are needed. Our brain, it seems, appears to spontaneously organize its networks into scale-free patterns whenever we engage in complex thoughts, only to disintegrate once those thoughts are disturbed.[21] And this kind of functioning seems to typify the operation of our brain. Earlier we encountered theta waves, just one of several regular oscillations of brain activity that correspond to particular mental states. Alpha waves, another example, are associated with drowsiness and wakeful rest. But there is also electrical activity in the brain that isn't regular or periodic. Instead, the background activity of the brain seems to be more like white noise, or the static interference common in radio-wave signals. Most researchers tend to strip out this kind of noise because it obscures the periodic signals of interest, and the assumption is that this noise is meaningless. But the statistical structure of this noise reveals, once again, a scale-free pattern, and this characteristic pattern may in fact reflect the relative state of consciousness of a person.[22] Patients in a coma, for example, may exhibit a different statistical pattern than those who are simply asleep.

A huge amount of research into the importance of this scale-invariant noise has been undertaken in a wide range of other scientific disciplines. In the 1980s, the Danish physicist Per Bak advanced the idea that this type of noise, with its distinctive statistical signature, arose from systems that were operating in a state close to a point of collapse—they were described as being "critical," in the same sense as is meant for nuclear reactors. In simple terms, a critical system is one that spawns a chain reaction in which each particle (in the nuclear example) triggers the liberation of another, which triggers another, and so on. A system close to criticality, therefore, is one that can rapidly trigger a chain reaction that results in a switch to a very different state. As long as the system remains below the critical threshold, therefore, it can maintain the status quo and continue to operate effectively. In fact, it is when it is close to the critical point—the so-called edge of chaos—that it operates *most* effectively. Studies show that the brain spends more time close to the critical transition during normal functioning than it does in either very ordered (subcritical) or very disordered (supercritical) states.[23] It's a delicate balancing act in which the brain attempts to maximize the amount of information that can be transmitted within it (a state close to criticality) while also maintaining stability of the neural networks needed to transmit the information (requiring a subcritical state). So again we discover not just regular oscillations in our brain waves, but a switching behavior, a flickering of our lantern, our guiding light, as it pushes the limits of cerebral connectivity and information processing, each time reaching the critical point and collapsing in a neuronal avalanche of electrical activity that resets the brain just like we saw in the pacemaker cells of the heart. We are a beacon, glimmering, flickering, switching, and trembling. But we are (a)light, and light is life.

Paying Attention

The simultaneous and inseparable wave-particle nature of light beautifully captures the indivisible unity of its combined physical-metaphysical associations, and the direct translation of that latter concept to the physical and nonphysical self. Both Plato and Descartes considered mind and body to be separate entities but defining where and how those "two modes of existence"[24] might be removed from one another has so far eluded satisfactory resolution. Instead we might see our self as a more complex and less easily definable phenomenon that exists both as a point in time—a particle, an element of physical matter—and simultaneously as an energetic wave that moves through space and through time. Light enables the twin aspects of our self to experience the world, and it is the light of understanding that allows us to make sense of it. We are primed, even before birth, to detect and respond to physical light. Even mice, born blind with eyes closed and unable to see the world around them, nonetheless prime their visual system with spontaneously generated retinal waves that mimic the sensory experience of freely moving around.[25] By presenting the brain with synthetic but structured impulses in this way, essential information is transmitted that allows the newborn mouse to self-organize its visual system in time for when its infant eyes finally observe the world. In humans, our visual system is operational from birth, but it only responds to a fraction of the electromagnetic spectrum, even when fully matured. As such, our concept of reality is based on only a fragment of what really exists,[26] so in order to build a seamless framework to enable both physical and nonphysical "us" to function effectively, we rely on interpretation. And the interpretations we make are guided to a large extent by the importance we attach to the fragments of information that our brain receives. Let's consider just one facet of light, color, to illustrate this point. In chapter 4 we saw how people in different parts of the world attach a different degree of salience to colors, depending on what is most meaningful to them or what is dominant in their landscape. To once more quote the artist Robert Henri, "colour is only beautiful when it means something,"[27] or Paul Cézanne, "colour is the place where our brain and the universe meet."[28] Physicists consider it somewhat less romantically, but the essential idea is the same: "There are no colours in the external world, independent of the process of perception" (Fritjof Capra);[29] "All colours are one thing, seen in different states of motion" (Frank Wilczek);[30] or perhaps most dryly of all, "colour is not a definite property of things in themselves, it is a useful device for the brain to keep track of objects" (neuroscientist Anil Seth).[31] Something as simple as our perception and interpretation of color, it turns out, is, in fact, complex and full of nuance.

Regardless of the source of the sensory information stream, those sensations would be meaningless in the absence of rules and structure imposed upon the data to contextualize them within the framework of our experiences and preconceptions. When we hear sounds such as speech, for example, we segment the incoming sensory stream into chunks to make processing easier. But how do we decide where to define the boundaries of a segment of speech? For that, we need to be able to predict, or at least anticipate, what is coming next.[32] In conversation, our pupils dilate synchronously with those we are talking to when we are most engaged in what they are saying, and we make eye contact as a way of signaling the peak of this mutual engagement.[33] Neural activity in the default mode network synchronizes between individuals according to the emotional content of the narrative.[34] And at the same time, we are processing the tone of the speech sounds in functional regions of the auditory cortex that are separate from the regions used to extract the linguistic meaning of that speech.[35] Yet despite the complexity of this multifaceted neurophysiological response, we also manage to continually evaluate the content of the conversation and update our interpretation of it through time. Through physiological synchronization and parallel processing of speech, we make a "best guess" as to where a natural boundary in the dialogue might occur and use this to define a segment that our brain can process. But there's a catch. Because in order to be able to make successful guesses and to process incoming speech efficiently, we need to not only be able to predict segment boundaries, but also be able to predict with some accuracy the content and future direction of the narrative. In slow-moving, highly predictable conversations, this isn't too much of a challenge and we can afford to digest the conversation in relatively large chunks. But when our partner's thoughts jump around chaotically, barraging us with a stream of barely connected ideas, or when we listen to a language in which we are only minimally proficient, the uncertainty of our predictions becomes so great that we are forced to process the conversation in much smaller segments in order to reduce the cognitive load. In both scenarios, the same feedback loop exists, but it dynamically adapts to the changing conditions. First, incoming sensory inputs are segmented and interpreted in line with existing knowledge. A prediction is made as to the meaning and relevance of the content being received, and this defines how the next segment of information is processed. This loop of sensory exposure and cognitive interpretation continually adapts as the conversation evolves, allowing our brain to optimize its processing strategy for the dynamic and potentially unpredictable content it is exposed to.

The benefit of accurate prediction, just as we saw with Neolithic societies in chapter 5, is that it allows us to adapt and optimize our strategy in a way that affords us some kind of advantage, a learning outcome that promotes greater success in whatever task we are dealing with. But it also minimizes our energy expenditure, allowing us to be more efficient. The better able we are to predict something, the less energy we need to invest in processing all the incoming sensory data—we just need enough to confirm or refute what we already know. In essence, we take what we can glean from our sensory inputs but fill in the rest based on prior knowledge. It is interesting to note how modern scientific understanding, based on neurological mapping and imaging of the human brain in action, closely echoes that of the Greek philosopher Plato, who maintained that what we observe from nature is merely an imperfect representation of a purer yet "hidden" truth. Because perceived reality (our sensory inputs) was, in Plato's view, merely an approximation of the real world, study and measurement of the distorted (visible) world could never be capable of revealing the truth of nature in its purest form. Any deeper understanding of the true nature of things could only come from introspection—that is, from our own predictions.

Our brain then, in dealing with the sensory firehose of the external world, builds an internal reality that is defined both by what is "out there" and also by what is "in here" in our memories. Immanuel Kant proposed that what we experience as reality, in terms of physical space and the dimension of time, is purely a construction of the mind, that the objects we perceive are merely "appearances." But memory allows us to link these appearances through space and time and to formulate a worldview that makes rational sense. Experiences that trigger discord with our lived experiences and memories demand greater cognitive resources to process. Yet some discord is not just useful, but actually desirable. Studies of aesthetic quality in the visual senses and musical appreciation in the auditory senses show that too much predictability can diminish our enjoyment, and that deviations from expectations—as long as they are not too extreme—are what distinguishes for us those pieces of art and music that give us greatest pleasure,[36] because they stretch our imagination and trigger neuronal avalanches that push us to learn and grow.

And perhaps we deliberately seek out such experiences. Faced with a continual onslaught of experiential data, our brains need to make decisions as to where to focus cognitive effort. And the way they achieve this has less to do with focusing our mental spotlight on the chosen element of interest and more to do with filtering out those other elements that are of lesser interest. Just as we saw in the restructuring of social networks in chapter 5, our brain pays attention by "unfollowing" the stimuli that, in that moment, it isn't as interested in. But here's the thing—this process of unfollowing, or filtering, isn't a hard "no"; it's more like a "not just now." Because our attention, just like our symbolic lighthouse, is a light that oscillates and switches, rapidly turning its attentional beam on and off so that other stimuli can be considered, just in case something more interesting or important comes up. Just as our ice age climate switched from one mode to the other, optimizing conditions on Earth for life to exploit new opportunities, so too does our brain switch from one mode to another, exploiting the perception-interpretation feedback loop to forever keep us optimally engaged with the world.

Individuality, Our Sense of Self, and Self-Generated Thought

So far in our discussion of the "Self" we have considered it mostly in terms of the indivisible unity of the mind-body system. But we have purposefully sidestepped the thorny question of how "self" relates to "individuality." From a young age we identify with our self through the name or names that we are given at birth. Yet those names can, themselves, direct the kind of person we become or the careers that we follow. People who don't like their name tend to have poorer psychological adjustment,[37] and those with names considered to be unpopular are more likely to end up getting involved in crime. Even the sound of our name can affect how others respond to us, with sonorant, or easily flowing, names (Rose or Anne, for example) being interpreted by others as reflecting that we are more agreeable than people with hard or abrupt-sounding names, such as Kirk or Kate. A study based on more than seven million individuals found that more unique or unusual names corresponded with more unique or unusual jobs;[38] another study showed that chief executives with uncommon names tend to adopt less conventional business strategies.[39] So the individuality of our names matters, but what, exactly, constitutes an individual?

In conventional thinking, an individual is a "thing," a physically separate entity that is bounded in space. "I" am distinct from "you," because we each occupy a defined and discrete fragment of physical space. But this is a very limited view that struggles with single entities that do not easily conform to such a rigid framework. Consider the great banyan tree of Kolkata, India, for example. With nearly four thousand individual trunks formed from aerial roots descending to the ground, the great banyan covers an area of nearly five acres but yet is still a single organism. Or what about the siphonophore? The majority of these marine organisms, of which 175 distinct species are so far known to exist, live in the deep sea and float or propel themselves through the water. Some emit light to attract prey. But an individual siphonophore is not an individual organism; it is a colonial organism composed of a number of specialized but individual multicellular animals called zooids. If we were to give names to siphonophores, therefore, would one name suffice for the whole colony or should each zooid be named uniquely? Colonial individuality is thus quite distinct from organismal individuality but no less valid. Equally, there is a third type of individuality that primarily arises from and depends on environmental conditions that keep certain compounds of reactions bounded within a discrete space. The earliest life may have existed in such a form, a delicate formulation in which metabolic processes were possible only as long as the containing environment was maintained. Thus "individuality" is more of a continuum than a singular concept, an idea that again resonates deeply with Eastern mysticism. In the *Tao Te Ching*, Lao Tzu describes the Tao (the "way" of life) as "something blurred and indistinct," and in Buddhism, the Atman, the essence of a being, is seen as a singular unifying awareness, without separation between beings. To researchers exploring the information theory of individuality, the best way to define an individual is to accept this blurriness, this indistinct separation between being, and to instead base "individuality" on how well an entity preserves its integrity through time, propagating "information from their past into their futures."[40]

Which brings us back to the concept of memory and the idea that the self arises from the integration of what we experience now with what we know from before. And it is interoception, our "sixth sense," which we encountered earlier, that enables us to parse the sensory stream of conscious awareness. Interoception allows us to distinguish feelings and emotions that derive from internal bodily changes from those that are responses to the external, social world we inhabit. Our sense of self, our individuality, is therefore a complex neuropsychological process that can fluctuate through time as we give greater or lesser weight to external stimuli, interoceptive signals, or memories of past experience. A balanced sense of self must be aware of all three, switching attention among them, as we saw in other contexts previously, and then using whatever blend of these ingredients is most useful in advancing our understanding.

Neurological studies have proposed that our "conceptual self," a mental construct that relies on analytical thinking, resides in the left hemisphere of our brain, the side most commonly associated with speech, language, rhythm, and abstract thought. On the other hand, the "integrative self," which brings together thoughts, moods, desires, and so on resides in the right hemisphere, the side typically engaged when processing emotions and melody.[41]* Reduced blood flow in the parietal lobes has been reported

---

* The idea from the 1960s that dominance in the left hemisphere makes a person more logical and that a dominant right hemisphere favors creativity is not supported by recent studies.

from subjects deep in meditation or prayer, and the temporary inhibition of the right parietal lobe in particular appears to generate an increased sense of transcendence, or selflessness, and an increased willingness to forgive. Correlates between neurological functioning and spirituality are explored in more detail in the next chapter, but for now it is enough to suggest that this region of the brain may be a key locus for our sense of self.

But as we have learned previously, brain regions rarely work in isolation. And so although parts of the parietal lobe may have been implicated in the establishment and maintenance of our sense of self, our ego, they function within one of the much larger resting-state networks we touched on previously, the default mode network. In addition to the inferior parietal lobe, the DMN also encompasses the medial prefrontal cortex, the posterior cingulate cortex/precuneus, the lateral temporal cortex, and the hippocampal formation.[42] When the DMN is active, these regions oscillate synchronously.[43] And the DMN is surprisingly active; in fact, it acquired the name "default" to reflect the fact that this network is active whenever we are not engaged in focused mental activity, in essence it represents the "resting state" of the brain.[44] Considerable research has provided us with a remarkably clear idea of the central role played by the DMN—despite only being formally defined in 2001—in many normal and disrupted patterns of cognitive behavior. The DMN is active when we let our mind wander, when we are daydreaming about the future, thinking about others or about ourselves, or reflecting on the past. It handles autobiographical information such as personal memories, facts about ourselves, or our emotional state. It allows us to contextualize the behavior of others or to evaluate social concepts and to make sense of narratives in the context of our past and future selves. But beyond these largely tangible concepts, the DMN also facilitates our comprehension of things less easily defined, such as the perception and appreciation of beauty, and through its role in self-generated thought, the DMN can be a source of creative insight and problem solving.[45] Critically, it is the fact that the DMN coordinates both aesthetic appreciation and our sense of self that enables great works of art to be so emotionally moving, resonating with us deeply in ways that leave us feeling touched by a masterpiece and personally connected to the artist behind it.[46]

The connectivity of the DMN becomes increasingly established through childhood, but in later life this connectivity starts to degrade. Diminished functionality in the DMN has been implicated in Alzheimer's disease and dementia, whereas lower DMN connectivity has been identified in people who have experienced long-term trauma or who suffer from post-traumatic stress disorder. Yet reduced DMN activity is also reported from long-term meditation practitioners, and reduced DMN synchronization has also been found in response to psychedelic drugs such as psilocybin or lysergic acid diethylamide (LSD). So what dictates the activity of the DMN, and why can similar changes arise from very different conditions and have very different effects? There is still much to learn about the DMN, but one simple explanation for this apparent paradox is that self-generated thoughts can be both beneficial and detrimental; it just depends on the content and context of those thoughts. The DMN is switched off, just like the attention network we saw earlier, by the salience network, which decides which incoming stimuli are significant and warrant attention. But what if we get stuck in the DMN, trapped in a world of self-generated thought from which no external stimuli are sufficiently distracting to break us free? Hyperconnectivity in the DMN and enhanced activity of the network manifests as obsessive rumination that persists even during focused tasks, when the DMN should be off. Dwelling on the past, particularly in the context of our narrative self, often results in depression. Worrying about the future, on the other hand, leads to anxiety. But the insidious aspect to these altered mental states is that our brains are plastic, meaning that they are able to continuously rewire their connections to better facilitate the neuronal activity taking place. A feedback takes place, then, in which depression and anxiety trap us into a heightened state of DMN activity, which in turn strengthens the neural connections of that network and prunes ("unfollows") the connections that we no longer use. Without intervention, episodic depression and anxiety become chronic and self-reinforcing.

Self-generated thought is, however, useful because it allows us to "think beyond the current situation"[47] and to assign meaning to incoming stimuli by illuminating them with the light of prior experience or to engage in more abstract feelings that allow us to make sense of the world. But when those thoughts are primarily negative, they can detrimentally impact both our physical and nonphysical selves. Sir Isaac Newton, one of the foremost scientific minds ever to have lived, excelled at being able to see things differently, to use his internal thinking to approach problems in unique and productive ways. But this also manifested in highly neurotic behavior, such as his obsessive concern for scientific superiority or tendency to brood over past failures, and ultimately his constant anxiety led to a nervous breakdown. Depression and other illnesses arising from suboptimal self-generated thought patterns are increasingly being recognized not just as mental illnesses, but as emotional illnesses that affect the entire body.[48] Given what we have learned so far about the body-brain connection, about interoception, and about the role of our own thoughts in brightening or dimming the light we put out into the world, it is encouraging to see holistic approaches to treatment increasingly being investigated. Strong magnetic fields have been shown to relieve anxiety in mice, whereas depleted levels of specific bacteria in our gut microbiota seem to be commonly associated with a wide range of psychiatric disorders,[49] suggesting that dietary interventions could also help. More sleep is advisable, because chronic sleep deficits are associated with increased mental health problems due to their effect on brain connectivity, particularly in the DMN. And because people with psychiatric issues are also more prone to having sleep problems, a positive feedback arises in which sufferers become locked in a constant state of insomnia and depression. People with depression also appear to have fewer mitochondria—the energy generators of cells—in their neurons, indicating that the nervous system as a whole is running on empty and that metabolic assistance might prove beneficial. Physical activity can help, not only by releasing mood-enhancing hormones, but also because it encourages neuroplasticity that can allow "cognitive rewiring" interventions—changing the way we think—to achieve adaptive results.[50] By contrast, the serotonin theory of depression, despite its lingering popularity, is increasingly rejected by the research community because the lack of empirical evidence in its favor suggests that the illness has roots far beyond a simple "chemical imbalance" in the brain.[51]

The apparent efficacy of these recent advances is encouraging, but they all largely separate the physical self from the nonphysical self—treating the particle and not the wave. To see ourselves with clarity, removed from the demands, expectations, and inner reflections of the social environment in which we live, requires more than these physical interventions, however. To harness the power of a genuinely mind-altering feedback that truly speaks to our sense of self, we need to embrace the "void and cloudless sky" and the clear light of illumination that can bring calm understanding and growth, adaptation and optimization. In the last few years an increasing number of clinical studies have experimented with psychedelic drugs, often in small quantities ("microdosing") that avoid hallucinations but are still thought sufficiently powerful to alter the brain state of patients. Laboratory-synthesized LSD, psilocybin from mushrooms, mescaline from cacti, and dimethyltryptamine, the active ingredient in the South American ayahuasca medicine, have all been experimented with for pharmacological purposes, but studies differ in their reports of effectiveness. Typically, the psychedelic experience is one of enhanced sensory experience, and perceptions of those stimuli that become distorted in time and space—experiences famously recounted in Aldous Huxley's 1954 book, *The Doors of Perception*, in which he described his experiments with mescaline. But it is other characteristics of psychedelics that make them potentially useful for the emotional problems arising from negatively skewed self-generated thought. Psychedelics have "mind-expanding" properties that blur the boundaries between "self" and "world," and it was just such traits that early researchers such as the Harvard professor of psychology Timothy Leary explored in controlled settings in the hope that the careful use of hallucinogens could offer real benefits to psychiatry. In a study in the early 1960s, Leary found that administration of psilocybin to a cohort of convicted criminals lowered their reoffending rates from 60 percent to 20 percent. Together with influential writers of the Beat Generation, such as Allen Ginsberg and Jack Kerouac, Leary actively promoted the idea that psychedelics could help people discover a higher state of consciousness, one that would be both personally and societally beneficial. Despite this optimism, or perhaps because of it, President Nixon considered Leary "the most dangerous man in America,"[*] and by the late 1960s, all such drugs were criminalized in the United States and further research became more or less impossible.

Renewed interest in recent years, however, has led to studies using psychedelic drugs to successfully treat a range of psychiatric problems. Ayahuasca, for example, has been found to be highly effective in relieving treatment-resistant depression, primarily through an indirect route in which the ayahuasca experience cultivates spirituality, which in turn transforms a patient's ability to process emotions.[52] The light of our inner world, therefore, is part of the experience-perception feedback loop described earlier, and as such it exists as a flame that needs tending and protecting. For if its light becomes too dim, our inner self is outshone by the dazzling glow of the external world, a floodlight that leaves us dazed and confused, trapped in a primal mode of fight, flight, or freeze.

Feedback of Fear and Other Emotions

Fear is essential for our survival. But it needs to balance perceived risk against the cost of inaction, that is, the cost of not attending to our fundamental needs. Deciding how much fear is appropriate is a task that doesn't just depend on the brain, but on bodily feedback transmitted to the brain via the vagus nerve.[53] Once a threat has been avoided, the fear response should subside. But extinguishing the fear response in the brain appears to

---

[*] www.nytimes.com/1996/06/01/us/timothy-leary-pied-piper-of-psychedelic-60-s-dies-at-75.html

depend on, among other things, the state of our gut microbiome. Mice with a healthy population of gut bacteria were better able to adapt to changing threats in their environment than those without a healthy microbiome. And surprisingly, when young mice without gut bacteria were given a healthy microbiome several weeks later, their threat response improved, but it was still significantly less effective than that of mice who'd had the full bacterial population from birth. If the same applies to humans, it suggests that our gut health and nutrition from birth play a critical role in establishing long-lasting behavioral responses to fear. And such responses get compounded through time, because the memory of fear leads to epigenetic changes in our bodies. Just as the Great Depression of the 1930s induced epigenetic changes in babies that accelerated their later-life aging,[54] exposure to past traumatic events changes the levels of enzymes in our body that influence how our genes are expressed and in doing so links prior trauma to later-life anxiety.[55] In chapters 2 and 3 we encountered the predator-prey relationship used in ecological modeling to simulate changes in populations of interacting species. But that simple approach assumed that species recovery would depend only on the availability of food and reduced predation. In light of the emerging insights we've just seen that describe the physiological memory of prior trauma, it's worth investigating whether the simple predator-prey feedback really does apply in the natural world or not. The fight, flight, or freeze response that characterizes most species' reactions to a predation threat is instantaneous—it is an innate survival mechanism. But in free-living species like the snowshoe hares of the Canadian Arctic, repeated encounters with predators such as lynx or coyote lead to persistent stress that changes the hare's brain chemistry— changes that are comparable to human cases of PTSD. Fear memories become engrained in the amygdala, a brain region with strong connections to the hippocampus, the region that forms memories of daily events. Trauma, it seems, reduces the ability of the hippocampus to grow new neurons, meaning that the fear memories of the amygdala aren't erased or replaced as they should be—an instance when "unfollowing" would be a good thing. PTSD therefore traps its victims in a state of permanent remembrance of fear. In the snowshoe hares this means that they eat less than they should and feed less to their young, so they are less healthy, less resilient, and their population recovers more slowly. The memory of fear, therefore, exerts a physiological and emotional feedback so strong that it passes from one generation to the next, potentially triggering disruptive impacts that can cascade throughout an entire ecosystem.[56]

Finding resilience in the face of such adversity is challenging, and the external nature of predation threats, war, socioeconomic deprivation, and so on means that our initial instinct is to prioritize the physical self over the nonphysical. The light of experience instead of the light of understanding. Yet surviving is different from thriving, and to do the latter, psychologists believe that what we need is an internal "locus of control," an inner light that gives courage, strength, and hope and enables acceptance. Feelings of love release oxytocin in the brain, which calms the cerebral fear center and facilitates greater social engagement, and drugs such as methylenedioxymethamphetamine, or MDMA, mimic the action of beta-endorphin, another brain chemical that reduces stress and produces sensations of love, affection, and empathy. Under the right circumstances these chemical changes can bring about a sense of greater connection with other people, a reduced focus on the self, and a reflective state of calm understanding akin to that related by those who meditate or frequently practice yoga. Our emotions then are the very real but nonphysical consequence of the full suite of stimuli that our physical self receives. Balancing external influences against interoceptive signals, our brain makes a "best guess" as to how our body should respond. Emotions are the mechanism by which the brain encourages the body to act, to adjust whatever needs adjusting in order to minimize the difference between the reality the brain has predicted and the reality indicated by the senses.[57] And so once again we might recognize that switching between alternate behaviors or ways of being could be the most beneficial way to maintain both a healthy self and a positive sense of self. Plato imagined the human soul in pursuit of love as divided into three parts: a flying charioteer pulled by two winged horses, one light and one dark, each pulling according to their differing motivations and desires. Regardless of whether our goal is love or not, it is learning to control those animals, to coerce them into both pulling together to carry our chariot where we want to go, that is the difficulty we must overcome. And, in Plato's view, the best way to achieve such an outcome is to alternate our behavior between action, when necessary, and inward thought the rest of the time.

Our journey into the self has revealed it as an inseparable wave-particle unity, a construction that is neither solely physical nor solely nonphysical. We experience the world one way but understand it another. The physical side of our self relies on a brain that continually and dynamically reorganizes itself to meet the cognitive demands imposed upon it. Feedbacks allow the activity of billions of neurons to self-organize in ways that optimize information transfer, a fragile networking arrangement that operates at the end of chaos and repeatedly collapses in avalanches of electrical activity that sweep across the brain. Powering all that mental activity, our heart tirelessly pulses oxygen and other essential nutrients around our bodies, relying on the self-organized activity of vast populations of cells in the pacemaker node that "integrate and fire" to deliver synchronized bursts of electrical energy to heart wall muscles. Connecting every corner of our physical self with the control center in our skull, the vagus nerve is the communicator that relays messages back and forth in a body-brain feedback loop, a control system always adapting and optimizing our functioning to most efficiently achieve our goals. And as those needs change, functional networks of disparate brain regions switch on and off, smaller networks spontaneously organize according to scale-free, fractal relationships, just like the variability of our heart rate or the frequencies of the background electrical static in our brain. To build a picture of the reality we inhabit at any given second, we make predictions drawn from past experience and combine them with sensory inputs that are filtered for greatest relevance, and in doing so establish for the nonphysical aspects of our selves a narrative that places us in the context of space and time, allowing us to relive the past and anticipate the future. From our narrative self, contextualized with other information, we construct a sense of self, a concept of what it means to be the illuminating, flickering, radiant being that is "us," and through trauma and adversity, we adapt, learn, and grow. In doing so we find ourselves poised, perhaps not at the edge of chaos, but at the doors of perception, tentatively peering into the "void and cloudless sky" that unites us not just within our two intimately entangled selves, but with all of existence and with the clear light that shines above our consciousness, somewhere out in the "transparent vacuum without circumference or center," somewhere unknown, in a place *Beyond*.

BEYOND   7

*All objects are as windows, through which the philosophic eye
looks into Infinitude itself.*

—THOMAS CARLYLE, 1831

I magine a tree. Picture it; paint a portrait in your mind. It stands in
a field, branches upraised, exultant, enraptured, open to the heavens
above and to the light that brings life to this complex dynamical system.
This tree stands strong, robed in leaves innumerable and structured with
such intent that its branching follows geometric laws of self-similarity, patterns repeating over and over from the largest limbs to the finest filaments,
a living fractal self-organized and optimized for what it is, for what it needs
to be.[1] Welcoming the air into which it grows, the tree opens its branches
wide, and as we study the subject of our sketch we see that this verdant
umbrella is not just that of one tree, but of two, grown not far apart. A
breeze might jostle the boughs of these woody neighbors, sending cascading waves of gentle disturbance through their collective canopy, but as the
ripples subside, they become once again still, proximal but apart. We live
in this visible world, the realm above ground where what we see is often
all that we perceive. Yet beneath our feet, the width of the tree's crown is
reflected in a hidden network of roots whose spread might be twice that
of the branches above, so that even for trees whose branches barely touch,
their roots entangle right to the base of one another's trunks. Early in their
life trees invest in structural root growth, affording them firm footing and
stability, resilience against the elements. But as they age, new growth is
prioritized for thinner threads, new rootlets that continue to weave their

intricate connections through the substrate that anchors them and allows them to persist. Bound in a symbiotic relationship, each tree root is often entwined with fungal filaments that penetrate or wrap around their woody cells. These mycorrhizae act as nodes through which sugars produced by the tree are made available to the fungi in exchange for water and nutrients captured from the soil by the network of delicate fungal threads. Through this partnership these diverse organisms help each other through life and depend on each other for survival. And so it is that as each tree matures and its mycorrhizal network expands and becomes ever more connected, so too do our neighboring trees become ever more entangled with one another, ever more a product not just of the energy and sustenance they harness through their own endeavors, but also of the life imparted to them by those with whom their roots become invisibly connected.

With experience and research spanning decades, forester-turned-scientist Suzanne Simard has contributed more to our understanding of the hidden world of trees than perhaps any other individual. Initially treated with resistance, even skepticism, her diligent and meticulous approach has been grounded in the recognition that trees are much more than isolated individuals; they are interconnected members of an active and nurturing social network. Analyzing the electrochemical flow of molecules in the mycorrhizal networks she mapped, Simard and her collaborators have revealed not just that fungal-arboreal symbiosis enabled better access to nutrients and to water, but also that, through chemical signaling along fungal threads, trees could communicate with one another. Deep-rooted trees share with shallower-rooted plants water brought up at nighttime; others send warnings of predation or transfer carbon to younger saplings whose growth could benefit from additional help. This forest-wide tree-to-tree coupling led Simard and her colleagues to conclude that mycorrhizal networks "provide avenues for feedbacks and cross-scale interactions that lead to self-organization and emergent properties"[2]—concepts that resonate with the examples of scale-invariance and optimizing feedbacks we've already seen in previous chapters. But perhaps the most remarkable finding of Simard's research was that the very oldest trees in a forest played the most important role of all, acting as the central hubs from which radiated outward uncountable threads of diverse fungal species. These ancients—the "mother trees"—build rich, multilayered networks over hundreds of years, which use chemical signaling, just like the neurotransmitters in our brains, to send and receive messages through the hidden subterranean realm to and from their younger family and the wider forest community. Having lived through climates that the young saplings have yet to experience and having survived the episodic ravages of pests and disease that most likely led to the demise of some of their weaker contemporaries, the mother trees impart a wisdom, an intelligence acquired through survival, that protects the forest and lends to it an inner strength absent from plantations of single generation trees.

Old-growth forests are quite literally "wired for healing,"[3] and, just as we saw with the human brain in chapter 6, the neural network of the forest strengthens, prunes, or reconfigures its myriad connections over time so that it can adapt to its changing environment and optimize its functioning for the demands that are placed upon it. There is an intelligence to any such learning system, but what about sentience—the ability to feel emotions like pain or joy—or consciousness—a higher-level awareness of the existence of both our internal and external environment? In our exploration of the self, we encountered the wave-particle unity of our simultaneous and inseparable physical and mental existence, a sense of self and inner wisdom that emerges from an ever-changing balance of internal and external stimuli. Perhaps we could see this in a forest community, individuals that respond not just to environmental stimuli but also communicate within and beyond themselves, prosocially supporting their kin. Is an individual tree conscious? Or does a tree only experience a state of collective consciousness once it is wired together with others through its mycorrhizal web? Our quest to discover what lies beyond our *self* must start here, with an investigation of what consciousness really is, how it might have emerged, and what purpose it serves.

The Origins of Consciousness

Throughout the preceding chapters we have encountered systems in which "oscillators"—components that repeatedly and predictably change their state through time—might gradually become entrained or synchronized either with one another (such as Huygens' clocks) or mutually with an external forcing (such as elements of the global climate system all responding to solar forcing). We've seen previously that synchronized physical activity between humans blurs the self-other boundary and makes us more empathetic and that watching emotionally charged films together produces synchronized neural activity. Physiological and neurological synchronization between a mother and her child strengthens their emotional bond and enhances an infant's sense of security, fosters learning, and promotes better health. As we experience external stimuli, our brains process parallel streams of information, each stream up- or down-voted by the salience network according to what it deems to be most relevant to us at any given moment. To effectively process these streams, our brains rely on the rapid coupling of discrete brain regions into functional networks whose activity is separated in space but synchronized in time to maximize the efficiency of information processing. Across the tens of billions of neurons in our brain, cell-level synchronization becomes one of the defining factors in how we think, feel, and respond to the world outside. Furthermore, it is the synchronization of our brain activity with that of others that is increasingly seen as a foundational phenomenon for consciousness.[4] Social interaction paved the way for the first evolution of speech and the emergence of a Stone Age human culture, but the fact that synchronized neural activity occurs during cooperative—but not competitive—tasks suggests that our capacity to connect with one another relies not just on the complexity of our gray matter, but on the specific timing of its operations.

Our brain predicts reality from the stimuli it receives and the context it lends to those stimuli based on what we have experienced before. We blend present with past and continually update our prediction, our "controlled hallucination,"[5] as new information arises. We sense emotions—feelings that motivate us to take action—when our mental predictions fall short and our experienced reality is not what we anticipated. Driven from the outside inward, our body responds to its external world and sends messages to our brain seeking resolution for the dissonance it feels. Just like the self-adjusting control systems in chapter 6, our emotions drive us to minimize the mismatch between our predicted situation and the one that actually unfolds. Emotions are our body's way of asking us to look inward and sense what it is that needs our attention.[6]

But if we are an isolated tree, a lone self, with canopy aloft and roots hidden below, we might only benefit from nourishment found within the halo of our insular, singular drip line. Our consciousness remains thin, fragile, and impoverished. By choosing to engage with others, however, we might exchange information, ideas, and interpretations, and through this exchange we collectively shape our now-shared experience, our innate "way of knowing." From the complex web of neurochemical transmissions surging through our cerebral forest, first synchronized across our own brain regions and then with the oscillations taking place in the brains of others, arises a shared hallucination, a perception of our immediate environment that is richer, more complete, and more objectively constructed than the one we built from our own singular stance. Through interpersonal connection we find our forest, and that forest gives us the true sense of our world, a reality perhaps not as pure and fundamental as the one Plato might have wished for, but one that is, in that moment, shared and agreed upon and accepted as fact.

We no longer treat consciousness as an individual phenomenon, then, but one that arises from the collective. We might not know, or be able to know, precisely when the ascendance of consciousness first transpired, but it seems likely that the level of consciousness we experience today arose increasingly rapidly once started—a positive feedback in which a critical phase transition took place that accelerated intelligence and learning to such a great degree that we found awareness of ourselves through interaction with one another. Somehow, the psychosocial feedbacks that took place as early hominins developed in mental complexity and cultural sophistication brought forth a rapidly connecting neural architecture that established the necessary hardware for consciousness. But the evolution of consciousness itself depended on functioning, not just structure. As we saw earlier in this book, the cognitive cycle of Fritjof Capra allows a structure to adapt, evolve, and optimize itself for a purpose, but to do this it needs an external drive, and for our burgeoning array of neural filaments, that drive was the ever-increasing flow of information. Surrounded by an environment continually bombarding our senses with multimodal information, our brains developed switching mechanisms to promote or suppress stimuli considered more or less important. But the information still came in, and by learning to make associations between those streams of information, we began to integrate that information into knowledge—a generalization that allowed more efficient storage of experiences, much like compression algorithms allow more efficient storage of digital images or sound. Mapping relations between stored objects, expanding our hidden web, our brains harnessed associative learning in a way that rapidly enhanced our ability to survive and prosper. Most likely this process started long before humans, with the origins of the first multicellular life described in chapter 2. Perhaps it was even the Cambrian explosion 541 million years ago that really marked the inception of the "unlimited associative learning" feedback that propelled us to where we are now.[7]

Wherever and whenever consciousness first arose, it did so from exploratory tendrils permeating the fertile soil of our brains, connecting regions that processed different types of information and making inferences from the integration of those different data. The progressive reduction of uncertainty surrounding the things we perceived and the increasing unity of each of those experiences led to integrated information that formed the basis of consciousness. Our consciousness is not, therefore, simply present or absent, but exists along a continuum, a progressive increase in awareness and understanding that affords to each living thing a greater or lesser degree of agency in how it shapes the reality it shares with all others.[8] To do this most effectively, each consciousness must learn to integrate past memories and future probabilities with present-moment sensory information[9] and to do so instantly and with appropriate balance, such that the messages given greatest weight in the decision-making process are those that offer the greatest benefit for survival. But how "instant" is *instant*? If we were to live entirely in the instant, we would be like the overly responsive control system of chapter 6, constantly switching on and off in a trembling state of anxiety. Instead, just like that system, our brain—our consciousness—relies on a "lag" that allows information to be processed and responses to be defined before we are even aware of anything happening. Our consciousness is a memory system, one that perceives parallel streams of stimuli but integrates and interprets those data before presenting to our awareness a picture, a prediction, of "reality."[10] This delay amounts to about half a second—the time taken for our unconscious brain to absorb, filter, and integrate incoming information and parcel it up in a way that our conscious brain can then "remember" it. Since conscious processes are generally too slow to be useful for any kind of activity that requires really rapid responses, it is our unconscious that takes care of most of our decision making and the actions that arise from those unconscious choices. Thinking (consciously) when we play a sport or a musical instrument impairs our performance, but relinquishing control to a "flow state" of the unconscious mind allows our cognitive performance to be everything it can be. We might even harness the benefits of the unconscious mind by asking a question then thinking about something else. The answer then comes quickly.[11]

The "information integration theory" of consciousness presents an elegant framework for establishing how neural structure and connectivity, driven by an external flow of sensory information, gave rise to a sliding scale of consciousness depending on the informational relationships within that structure.[12] And just as with Capra's cognitive cycle, the depth and breadth of our consciousness can evolve through learning, harnessing the plasticity of the brain to strengthen, prune, or completely rewire our brain regions depending on what is needed to process the stimuli we are exposed to. Feedbacks ensure that the inward flow of information triggers neuronal growth that in turn enables more efficient processing of that incoming information. This frees up other resources and facilitates further growth to better entrain other stimuli or to associate new experiences with past memories in ways that allow new knowledge to emerge. Importantly, one of the main implications of the information integration theory is that "any mechanism that has properties of information and integration, whether a brain or not, whether biological or not, will have consciousness."[13] This startlingly profound inference has wide-reaching implications for how we relate not only to one another, but to all other life and to all the other dynamical systems we might observe to exhibit characteristics of this kind of consciousness.

Through consciousness we are no longer just a point, a particle, an isolated tree, but instead a rich and diverse forest, an energetic field, and a multidimensional wave. We are each a star in a constellation of possibilities, sharing our field of consciousness with all other conscious beings. There are specific wavelengths of brain activity, such as alpha waves, to which our brains are most sensitively tuned and that provide a mechanism for brain-to-brain synchronization. Through the time-synchronized activity of our neurons and the ability of this synchronization to operate between individuals, we now know that consciousness is a nonlocal phenomenon, meaning that simultaneous changes happen despite physical separation in precisely the same way that entangled quantum particles share a connection that persists even when isolated from one another.

Riders on the Storm

For a species that has worked hard, evolutionarily speaking, to attain a state of self-aware consciousness, we are remarkably keen to find ways in which we might now change that state to temporarily experience a reality that is perhaps more uniquely ours and less shared with the world at large. Our brains host a network of structures that collectively form the mesocorticolimbic circuit—the "reward system." This network connects the cortex, the basal ganglia, and the thalamus back to the cortex in a loop that responds to rewarding stimuli by way of neurotransmitters like dopamine, gamma-aminobutyric acid (GABA), or glutamate. In its simplest form, the reward circuit encourages behaviors that promote survival, such as seeking energy-dense food or prospective mates for reproduction. By offering "rewards," often in the form of a feeling of pleasure, this network preferentially conditions us to invest time and energy in things that will promote survival both of ourselves and of our species. This feedback is thus one of reinforcement learning, wherein an action triggers the release of a neurotransmitter such as dopamine, which makes us feel good and encourages us to "want" to undertake that action again in order to repeat the pleasurable feeling it produces. Perhaps one of the reasons song became such a key part of interpersonal connection and the emergence of human culture, as we discovered in chapter 5, is that music is one of the sensory experiences that triggers the reward circuit. When we listen to music, dopamine signals from the midbrain produce sensations of pleasure at two distinct phases of the reward cycle. First, there is a pleasure response in anticipation of harmonic changes or changes in loudness, followed by a second pulse when that pleasurable change peaks.[14] But once again we come back to the concept of prediction. As we saw in chapter 6, music that is too predictable doesn't invoke the same degree of pleasure as music that is more complex or harder to predict. One reason for this, then, is that one part of the "reward," or source of listening pleasure, is in the anticipation of something that we have predicted, while the other part lies in finding out that we were correct. We like to be challenged, but we also like to be right.

Our ability to make accurate predictions for music that is challenging and unfamiliar depends on the quantity and breadth of our past musical exposure and our capacity to retrieve and integrate memories of those past experiences with what we are now hearing. Given the attractive nature of the dopamine pleasure response, it is no surprise that we are willing to invest considerable cognitive and energetic effort in long-term musical training, a pursuit that leads to persistent structural and functional connectivity in our brain.[15] Music and dance, through carefully coordinated and synchronous activity either by an individual performer or between performers and their audience, also enable neural rhythms to become entrained, and because focus is required, attention to other thought drops off. We become absorbed in what we do, immersed in the unconscious flow state and rewarded with dopamine as our consciousness predicts and adapts its perception of the evolving reality we share with one another. Rhythm, synchrony, pleasure. A simple yet effective feedback that keeps us doing what feels good, and it is through this same feedback that arises another fundamental human response—our lust for sex. Sexual pleasure is one of the most intensely rewarding experiences humans can experience, and its pleasurable physical feedbacks drive us to pursue it for much of our life. Through sexual activity we benefit from enhanced cognitive function as the medial prefrontal cortex rapidly builds new dendritic spines across its neurons, helping us to more flexibly switch between different mental tasks. During sex we consciously control our behavior to generate steady rhythmic motion and switch our attention networks to the here and now—the more we attend, the greater the pleasure, and the more we want to attend.[16] Intimate pair-bonding feels good because our state of consciousness changes; we achieve a state of enhanced neural synchrony not just within ourselves, but with our partners. Theta rhythms in our brains gradually build during intercourse, rising steadily just like the relaxation oscillator systems we have seen previously. Experiencing a diffusion of focused attention, we sense a loss of ego, we dissolve into the other, entraining our bodies, hearts, and minds in a harmony that triggers heightened perceptual vividness and increased emotional intensity. We experience a glimpse of the "void and cloudless sky,"[17] the infinitude of the heavens and the clarity of transcendent awareness. And just as with our musical predictions, we enjoy the anticipatory phase as much as the climax because the reward circuit is triggered both during the expectation of what is to come and then, finally, at its peak.

Whereas activities such as those above affect the reward circuit, pleasure, and emotional aspects of our consciousness, there are other indulgences that expand our consciousness in other ways. In chapter 6 we saw how the brain is continuously barraged by stimuli from the external world, from within our own bodies, and by thoughts bubbling up in our minds when we let the default mode network ruminate without control. To make sense of this onslaught of information, the salience network switches our attention from what isn't most relevant to what is. By filtering out the noise, our attention remains where it needs to be to keep us alive. But what if that filter weren't there? Are there ways that we could allow our consciousness to experience the totality of our sensory inputs and our unconscious thoughts? Ways of knowing that, to echo E. F. Schumacher once more, might bring us greater understanding, not just greater knowledge?

Plants with a range of intoxicating properties have been cultivated and used since the Neolithic, suggesting that our desire to experience altered states of consciousness goes back at least five to ten thousand years and most likely much further.[18] For many tribal groups, the gathering of special plants, such as the mescaline-containing psychoactive cacti of Mexico, was a spiritual or religious pilgrimage accompanied by much ceremony and a reverence for the power of the plant. Over recent decades, the list of researched psychoactive compounds has steadily grown and the increasing interest in their potential medicinal properties has led to an increasingly quantitative scientific exploration of their effects. In a recent review of nearly seven thousand descriptions of hallucinogenic experiences spanning twenty-seven different drugs, researchers sought to identify the key commonalities of the altered states of consciousness that each participant had experienced.[19] The leading descriptors retrieved from fourteen thousand keywords pertained to ideas of mental expansion, an awareness of Earth, reality, the universe, and existence and also concepts of liminal or mystical beings—spirits, entities, or aliens. Frequently the psychedelic experience is also associated with the interconnection of these concepts, a reduced sense of individual identity and a greater propensity to feel part of a larger *something*, a diminished ego that is more intrinsically embedded in the world and a connection to a universal wisdom that is embodied in the self and in nature. To neuroscientists like Anil Seth, the psychedelic state is one that reflects a "change in overall conscious level," a shift along the continuum of information integration and complexity.[20] Our hallucinogenic consciousness is one whose delicate hyphae are suddenly rampant in the forest of awareness, accelerating its learning as node after node entwine to trace out an invisible labyrinth of ideas and possibilities from a configuration space instantly infinite and bursting with potential.

Although the perceived experiences under psychedelics differ from one person to the next, the patterns of neural adjustment they trigger in our brain are becoming increasingly well mapped. Compounds like LSD, psilocybin, mescaline, and dimethyltryptamine (DMT) produce altered perceptions of self and the world by modulating the serotonin system in the brain, the network responsible for feelings of enduring contentment rather than transient pleasure.[21] By binding to the 5-HT$_2$A receptor, these compounds weaken the organizational control of the default mode network (DMN), which normally regulates brain activity of a suite of subordinate neural networks.[22] Brain regions rich in 5-HT$_2$A serotonin receptors become more functionally interconnected, while areas controlling spatial navigation and locational awareness exhibit decreased connectivity, perhaps explaining reported feelings of ego dissolution as we lose our sense of space.[23] Other neurophysiological changes are localized: an increase in blood flow takes place in regions of the brain responsible for processing visual stimuli, correlating with the intensity of visual hallucinations reported. Immediate and long-lasting growth of dendritic spines of neurons in the medial frontal cortex (important for mood and cognition) rapidly increase the number of synapses sensitive to the pleasure-giving reward circuit neurotransmitter glutamate.[24] As research continues, we will undoubtedly uncover more details of the transformations and adaptations taking place, but what is already clear is that the brain as a whole becomes less modular: the psychedelic brain is a more integrated brain,[25] an expanding web of virgin connections and novel associations that enhances our cognitive abilities and our conscious experience in ways that persist long after the intoxicant effects of the compounds have disbursed.

So much for the effects, but what of the mechanisms behind these mind-expanding experiences? "If the doors of perception were cleansed," wrote William Blake, "everything would appear to man as it is, infinite. For man has closed himself up, till he sees all things through narrow chinks of his cavern."* How then do psychedelics cleanse the doors of perception and widen those chinks enough to change the balance of feedbacks that normally acts as the gatekeeper of our daily conscious state? Piecing together the quantitative experimental evidence, a picture emerges that aligns with what we have already learned about "normal" brain functioning. And it seems to all come down, once again, to the concept of prediction. Previously we saw that, by making experience-informed predictions, our unconscious brain can guide our conscious actions in ways that, ideally, give us the greatest chance of survival and the optimal state of being with respect to the environment we find ourselves in. But to do this, our brain relies on feedbacks that either enhance or suppress individual streams of incoming information—the control system that maintains our status quo. Psychedelics disrupt these feedbacks, such that the higher brain regions that control associative cognition are less able to filter or selectively suppress the torrent of sensory signals bubbling up from beneath. Changes in the coupling of our large-scale brain networks like the DMN mean that the predictions made by our higher-level brain regions are less guided by external stimuli, leading to a breakdown in the rigorous "fact-checking" and data assimilation that our unconscious brain usually performs. With less accurate predictions of what the outside world really looks like combined with an overwhelming flow of unfiltered sensations, the dissolution of our sense of self, and a distortion of our sense of space and time, we experience a "disintegration of ordinary self-awareness"[26]—and instead gain an expanded perception of the infinitude of existence.

Breaking this tightly regulated sense of self is why psychedelics appear to show promise for chronic and drug-resistant emotional disorders such as neuroticism, depression, anxiety, and post-traumatic stress disorder. More widespread brain integration through psychedelics counters many of the negative rewiring effects of the emotionally damaged brain discussed in chapter 6—rewiring that progressively reinforces negative thought patterns as linkages between diverse brain regions are pruned and local connections within a single region are strengthened. Just as the forest depends for its health on the diversity of its interspecies and intergenerational bonds via fungal filaments that foster sharing, protection, and resilience, so too does our neural community depend on the two-way conversations that convey information and confer wisdom. To some, the state of consciousness reached

---

* William Blake. *The Marriage of Heaven and Hell* (1793).

under psychedelics mirrors that achieved by experienced meditators: a state
of heightened awareness, of enlightenment. Evidenced by similarities in
brain waves measured by electroencephalogram (EEG), a key similarity
appears to exist in the coherence of the signal and in the functional neural
connectivity. No wonder then that shamanic use of psychedelics has been
so popular for millennia. Sensing the spirits through ayahuasca ceremonies
or ritual mescaline consumption, Indigenous cultures found connection to
their inner self, to their community, and to their world, and they saw this as
essential to their emotional and physical health.[27] Expanding our appreciation of what it means to be conscious clearly improves our own situation,
but it goes much further. Because through experiences such as these we
begin to see more clearly that, if *we* are able to experience consciousness in
more than one form, then perhaps we might see that all living things possess some form of consciousness, different from our own, but on the same
sliding scale of awareness.[28]

Star-Struck

In a forest we can count the trees; we can measure the girth of trunk or
radius of crown; we can calculate and classify and derive laws to predict
their growth. This is the visible world, and this is one way of knowing.
But the less-visible world is demonstrably just as real, though it remains
hidden. What we don't see are the webs intricately woven through the
soil, entwining a multitude of species of fungi, grasses, shrubs, and trees.
We don't observe the electrochemical signaling they employ to communicate with one another, the social networks established over centuries and
continually adapted to bestow resilience and survival on their hosts. Except
to those who study them, these pathways are less real to most of us than
the countless stars in the night sky and the imaginary constellations we
trace between those distant points of light. Yet they are there, and without
them our forest would die. Learning to see this hidden world as a complex
dynamical system that operates through feedback-driven cognitive cycles
of learning and adaptation that evolves an intelligence and consciousness
of its own is a second way of knowing. *"There are hundreds of ways to kneel
and kiss the ground."**

In evolving to a state of self-aware consciousness, our species became
able to reflect on its own emotions, its past experiences, and to seek explanation, context, and purpose. We feel we need knowledge, but what we

---

* Jalāl al-Dīn Muḥammad Rūmī. *Rumi: The Book of Love: Poems of Ecstasy and Longing*, trans. Cole-
man Barks (New York: HarperCollins, 2005).

really need is understanding. What is seen is not the whole of reality, but to see that totality, we must extend our faculties beyond the filtered subset of information that our salience network allows to permeate through to our thinking, conscious mind. How did we first come to realize that our visible world was not the full story? Certainly we may have made use of intoxicating plants, an accidental foraging discovery retained and shared for its beneficial uses. Other theories abound. Through the neurological transformations that take place during intimacy and intercourse and the reward mechanisms in place to ensure repeatability, mammalian sexuality opened a window into an otherwise barely reachable place, a transcendent experience in which a temporary loss of our sense of self became comingled with a feeling of a unity with nature and a special connection with something larger than ourselves—the cosmos, perhaps, or a state of pure awareness. When we make love, the right amygdala and left fusiform gyrus become deactivated, lowering our innate fear and aggression responses, while a similar deactivation of the anterior orbitofrontal cortex at the point of orgasm inhibits our decision making and leaves us immersed in the moment. Alpha wave activity in the right parietal lobe, the area responsible for our sense of self, transitions to delta wave activity more typically associated with slow-wave sleep, while the release of oxytocin and vasopressin promotes altruism, empathy, and a positive focus on others.[29] Something as simple and habitual as being intimate with a partner may, it seems, have been one of the ways that we first sensed a reality that extended beyond the limits of our own blinkered existence.

But sex was surely not the only way. Some argue that it was feelings of awe that became the precursors to spiritual awareness, feelings brought on by complex and information-rich stimuli such as music, art, or landscapes.[30] To early hominins, the experience of something overwhelming, something difficult to assimilate into their existing understanding—a volcanic eruption, earthquake, or solar eclipse perhaps—pushed them to modify their beliefs to accommodate the new information. A sense of awe goes beyond that of transient surprise: it is associated with a diminished sense of ego, a dissolution of the sense of self; it triggers a change in attentional focus and memory and results in a long-lasting change in belief. Coping with awe means struggling to understand something for which we have no prior explanation, and through that struggle, we form new ways of knowing, of meaning making. Perhaps a feedback arises in which we ascribe each new experience of awe to the same source, a confirmation bias that self-strengthens through time to become real.

Our spiritual moments are "those moments when we feel most intensely alive,"[31] a fullness of mind and body and a connection to other living beings. Hunter-gatherers of the /Xam group, part of the San population of southern Africa, revere their relationships with other animals so highly that they believe "threads or cords or beams of light" link living things to one another as well as to their god.[32] *We are a beacon.* The /Xam have a different way of knowing compared to Western science, but they are not confused or ignorant, nor are they alone. Indigenous cultures on the other side of the world, in South America, believe that they, with the aid of ayahuasca, can see "spirits," the light emitted from every living being.[33] Our DNA is the source of this ultra-weak, highly coherent, visible light.[34] We might not see it, but it is there, and as we saw in chapter 6, it can be measured.[35] The coherence of this laser gives a pronounced sense of color, luminescence, and depth, qualities strikingly similar to reports of the visual aspects of psychedelic experiences. Regardless of whether or not we can see the electromagnetic connections between us, some researchers suggest that it is the ties linking one to another that underpin the biological basis for our spiritual beliefs. These ties might manifest in our capacity for empathy and our recognition of empathy in others. Sensing compassion from our companions, we align our behavior with theirs, and those mutual changes strengthen the bond between us. As the bonds between individuals grow to pervade a group, we refer to the emerging sense of unity as "community spirit," a sense of connectedness, support, and mutual security.[36] The "spirit" of our spiritual beliefs, therefore, might lie here, in the connections between people, between beings, in the spaces between the reality we perceive ourselves and the reality experienced by the life with whom we share the singular and continuous field of a collective and integrated consciousness.

"We are connected by invisible links" asserted the physicist, inventor, and visionary scientist Nikola Tesla. Critical of psychic and spiritual phenomena, Tesla nonetheless recognized the overwhelming electromagnetic evidence for the reality of our existence: "we are automata entirely controlled by the forces of the medium being tossed about like corks on the surface of the water." Given this situation, Tesla felt that each of us was only a part of something much larger and that, throughout our lives, we remain forever connected. "Though free to think and act, we are held together, like the stars in the firmament, with ties inseparable."[37]

How then might we see what can't be seen, the invisible essence of a larger reality hidden from our normal conscious awareness? Sex, drugs, and rock 'n' roll may offer glimpses into this beatific world, whereas awe, empathy, and ritual may help us recognize and accommodate the wider understanding it brings, but how do we evidence that which lurks beyond our senses and make meaning of something for which only fragments of insight exist?

Perhaps it's an impossible task, a fool's errand. But we live, daily, with a suite of accepted truths for which we, personally, have no direct evidence or understanding. We accept gravity not because we can see this fundamental force, but because we witness the consequences of its action. Invoking a force called gravity makes it easier to understand why an apple falls from a tree or why our Earth orbits the sun. We accept the explanation of gravity because it readily explains the vast majority of observations. But it remains an inference, and one that isn't as straightforward as we might like to think. Newton believed that gravitational attraction depended on the mass of an object, but the laws of Newtonian gravity didn't work for the relativistic universe of Albert Einstein. A modification was therefore needed—Einstein's general theory of relativity. Einstein proposed that gravitational attraction depended on energy, of which mass might be just one form. For Einstein, all forms of energy had gravity, and this gravity pushes rather than pulls. But both of these ideas break down when gravity becomes very weak, such as at the far edges of galaxies. To better explain the motion of faraway stars in those extreme circumstances, it turns out that a different interpretation of gravity is needed—one called Milgromian gravity.[38] And at the other end of the spectrum, there is a suggestion that we need yet another theory, quantum gravity, to understand how gravity behaves either at very small scales or at very high energies.

The fact that we need at least three and probably four different theories to explain something we can't even see isn't to suggest that none of these ideas is correct. In fact, they are all correct, when applied in the most appropriate context. But there is a cautionary tale here. Had we stopped trying to explain the things that didn't fit with the first theory, we wouldn't have found the second, and without questioning the second, we wouldn't have discovered the third, and so on. And that's how it is with consciousness. It was once thought that humans occupied a unique and special place in the evolutionary scheme of things, the only animals to have feelings and to be self-aware. Yet continued scientific investigation now shows without doubt that many animals, even crabs, are sentient, that several mammals are self-aware, and that consciousness itself is not simply a singular state but rather a sliding scale of awareness. Since this scale extends beyond the level of consciousness typically employed in our day-to-day functioning, it begs the question, why do we have it?

If the filtered subset of experiences that our unconscious allows to permeate through to our self-aware consciousness is sufficient to ensure our survival, what purpose is there for the perception of a wider reality? Is it all about security, species persistence, and procreation? Or is there more to life than simply our survival—an innate drive to not just know but to understand our existence? In his *Symposium*, Plato proposed that through a love of beauty we might move progressively upward in a spiral of intellectual expansion that moves from knowledge to understanding and culminates with recognition of the ultimate reality: "What else could make life worth living, my dear Socrates . . . than seeing true beauty?"[39]

Where does this ultimate reality, this true beauty, fit into our conception of consciousness? More than 80 percent of the global population consider themselves religious or spiritual. Although religious belief is concerned with unified systems of belief and practices or ritual, spiritual acceptance is more agnostic and more reflects an enduring recognition of forces that cannot, within our current framework of knowledge, be objectively proven.[40] Somewhat remarkably, the relative fervor of these beliefs might depend on the connectivity and functioning of key brain regions such as the parietal lobe—the area we saw earlier that was down-regulated during sex. Patients with damage to this frontal lobe reported increased spirituality or religiosity, but as we have seen previously, such associations are more likely tied to discrete brain circuits, or networks, rather than to any one particular region. Investigating data from nearly one hundred patients who had undergone surgery to remove a brain tumor, researchers found that their self-reported sense of spirituality was different after the operation than it was before. But in these cases it wasn't the parietal lobe that was affected, but an area called the periaqueductal gray (PAG). The PAG itself is one of the oldest structures in the mammalian brain and is highly associated with fear conditioning, the modulation of pain, and feelings of altruism and unconditional love. When positive nodes of this brain stem region were surgically removed, the researchers found that it led to a reduction in a patient's spiritual beliefs, whereas operations that impacted the negative nodes of the PAG led to an increase.

With evidence that specific regions and networks are intimately associated with our perception of something beyond ourselves, it seems that our brain is structured for spiritual awareness. Whether or not we engage and develop that ability partly depends on our neurological connectivity and partly on our intentional practice. In this and earlier chapters we have seen that the default mode network processes self-referential thought, whereas the parietal lobe plays a key role in the processing of sensory inputs. The amygdala controls our emotions, particularly fear and aggression, as well as decision making, while the salience network is the critical switch that decides what messages need our conscious attention and which do not. When things go wrong, however, this normal pattern of functioning breaks down. Hyperconnectivity of the DMN, for example, is routinely associated with depression, whereas disrupted behavior of the hippocampus prevents fear memories in the amygdala from being successfully erased, leading to long-lasting post-traumatic stress. And this is where, perhaps, spirituality can help. Clinical psychologist Dr. Lisa Miller has spent decades researching the relationship between the functional connectivity of disparate brain regions, emotional health problems such as depression, and the tendency of patients to adopt or develop spiritual beliefs. Her evidence from years of clinical and laboratory-based investigations suggests that reliving spiritual narratives quiets the overactive DMN and subdues the parietal lobe, just as sex does. But the parietal doesn't shut down entirely; it just reduces to a pulsing activity, and it is from this region that alpha waves emerge that, in empathetically connected partnerships, allow entrainment and synchronization of neural activity.[41] By harnessing the distinct neurological changes that take place in the spiritually awakened brain, we optimize our health and social functioning in ways that go far beyond simply surviving and instead enrich our lives in less quantifiable ways—increasing our sense of connectedness, our feelings of worth and purpose and of the intrinsic value of existence. "You are not a human being having a spiritual experience. You are a spiritual being having a human experience."[42*]

We evolved brain regions that collectively perform spiritual roles, perhaps as a way to help us go beyond mere physical survival. That such functionality remained beneficial is evidenced by the fact that our spiritual leaning is almost 30 percent heritable and that for spiritually aware children whose mothers are similarly inclined, the chance of developing depression is 80 percent reduced.[43] On the basis of this and other evidence, Miller argues that depression itself perhaps reflects a "spiritual hunger" and that emotional suffering encourages us to activate our network of spiritual perception as a way to better protect our self from future trauma. Just as the reward circuit we encountered earlier guides us to adopt behaviors optimal for our survival, so too, it seems, does the belief circuit. Both of these networks exert a reinforcing feedback that self-strengthens over time, fine-tuning our behavior and optimizing our functioning for the conscious reality we choose to inhabit.

---

* A similar statement is commonly attributed to the French scientist, philosopher, and theologian Pierre Teilhard de Chardin, but no robust evidence of such authorship exists.

Harnessing the Human Spirit

However we wish to conceptualize the meaning of human spirituality, whether it be in the form of a single figurehead and a collection of ritual activities, or a more diffuse and ineffable quality that permeates all of nature, or simply as something within us—an intrinsically personal way of knowing—there are common ways to embrace it and to explore it more deeply. The practice of meditation most likely emerged from Indian Ayurvedic medicine, with uncertain origins that go back at least to the first written descriptions from Hindu sources dated around thirty-five hundred years ago, and perhaps as far back as seven thousand years ago. The philosopher and mystic Patanjali, living sometime between the second and fourth centuries BCE, brought together the *Yoga Sutra*, a collection of texts describing the practice of meditation. The techniques of yoga were then brought to the West in the late nineteenth century by Swami Vivekananda, followed subsequently by the import of Transcendental Meditation by Maharishi Mahesh Yogi in the mid-twentieth century.[44] Yoga was originally concerned with the stillness of mind, to which later became attached the control of body through the physical poses we most commonly associate with yoga practice today. Through settling the mind, both yoga and meditation aim to cultivate well-being as well as insight into the true nature of all things.[45] Many forms of yoga and meditation have evolved since its origins, but from at least the first millennium BCE, the principal goal of the practice was the achievement of a mental state in which all outward distractions were eliminated, and the practitioner was left aware only of being aware, that is, with a state of pure consciousness.[46] This is a state in which experience has no content, no subject, and no object, a relaxed state of wakefulness in which the distinction between self and other is removed.[47] To Buddhists, this is a state of *suññatā*, or emptiness, and in this state thoughts emerge from, and sink back into, pure awareness. Transcending the noise of the conscious mind to achieve this state requires the observing of thoughts with detachment, turning the attention inward until only the subtlest levels of thought are perceived. And then, by eventually transcending even those most delicate of intrusions, we reach the blissful, superconscious state.

Among the most popular of today's countless forms of yoga and meditation are mindfulness practices such as the Tibetan *samatha*, which aims to calm the mind through focused concentration and is often coupled with *vipassanā*, the quest for insight and inner wisdom that can only arise from a quieted mind. And this quieted mind expresses itself physiologically in ways that can be monitored and measured. Studies of experienced meditators reveal reduced oxygen consumption, a lowered heart rate, lower skin conductivity, and an increase in both the density and amplitude of cerebral alpha waves.[48] The default mode network becomes less active during mindfulness meditation but more extensively coupled with other regions of the brain—analogous to the globally integrated psychedelic brain we saw before. Functional reorganization of the brain during meditation appears to enable the brain to become more aware of typical DMN activities—mind wandering and self-referential thoughts—and to keep these in check through enhanced attention arising from a stronger coupling to the prefrontal cortex.[49] Mindfulness meditation therefore exerts a three-way control, regulating attention, emotion, and self and body awareness, the latter modulated by functional changes to the insula, the brain region responsible for perception of the body's internal operation.[50]

Regular meditation reduces stress, anxiety, and depression not just because it is relaxing, but also because it rewires the functional networks of our cerebral neurons. By reducing activity in the DMN and parietal lobe, we become more empathetic and compassionate, less self-absorbed, and more receptive to the idea that we exist as part of a larger consciousness. But there are other meditation-induced adaptations that support our physical and emotional well-being. By diminishing our sense of self, we perceive less pain. This is not because we experience reduced pain—the stimuli are still transmitted to the brain—it is because we feel less of a personal association with that pain. In an fMRI-based clinical trial, participants reported a more than 30 percent reduction in pain intensity or unpleasantness arising from reduced synchronization between the thalamus and the DMN.[51] In other studies, researchers investigated genetic expression and immune system response of experienced meditators from nearly four hundred blood samples taken at different times during an eight-day intensive meditation retreat. Results showed enhanced immune functioning following the period of meditation and an altered expression of several genes related to oxidative stress as well as the genes governing the growth and death of cells.[52] Through inner engineering—hacking the invisible world inside us—we not only open ourselves up to a more complete understanding of reality, but gain measurable benefits for our physical, visible self as well.

So far, so good. But there's a problem. As we have seen, the weight of emerging research increasingly argues for a sliding scale of consciousness, a property that furtively crept into existence in its simplest form early in the evolution of multicellular life. As our neural anatomy continued to grow, the level of consciousness that accompanied it became ever more advanced, eventually reaching a point where we transcended the state of simply being "self-aware"—aware of our physical self—and achieved a knowing state in which we became aware of our own thoughts, meaning that we could then *think about our thoughts*. And because our consciousness evolved through social interaction, we also became aware that other people also had thoughts and that they might also be thinking about *our* thoughts. From that point on, it became increasingly important for us to not only respond to stimuli from our physical environment in order to stay alive, but also for us to correctly predict the thoughts and future actions of the people around us. The amount of information that it became necessary for us to process in this newly self-and-other aware conscious state drove the accelerated structural growth of our brain and the rapid increase in neuronal density that, as we saw in chapter 6, markedly distinguishes us from other mammals. But here's the paradox. If the evolution away from purely reactionary, unconscious functioning, toward deliberate, conscious functioning was so beneficial, then why is it that, through psychedelics or meditation or any of the other "mind-expanding" techniques we use, we find ourselves increasingly seeking a stronger reconnection with our unconscious mind?

To resolve this quandary, we might recap some of the key insights from the research we covered previously. The accelerated structural growth of our brains occurred rapidly (at least in evolutionary terms) and gave rise to a mass of gray matter that could be continually and instantly rewired and utilized for novel processing tasks and learning. Cerebral growth gave us the canvas for our portrait, but it was up to us to paint the picture. Depending on the specifics of our environment, the challenges we face, and the social interactions we have, we now weave a web of interconnected neurons that build atop our structural canvas layer upon layer of relationships, synergies, and interactions. The neural networks we develop are unlimited, adaptable, and instantly reconfigurable. We have infinite capacity for associative learning and for the continued integration and recombination of the information we absorb. And from what we have seen in this chapter, one of the most effective ways for us to enhance this kind of connectivity, to establish a globally integrated and less modular brain, is to tap into our unconscious not by thinking more intensely, but by perceiving more widely. When we focus less on the acquisition of knowledge and more on the understanding of what we already perceive, we discover that our self is not a singular thing, but is a part of a larger self, embedded in a consensus reality that embraces the full spectrum of a universal conscious field.

The conscious mind evolved, therefore, to help us parse the exploding volume of sensory information that our physically and socially aware brains began to receive. Acting as an executive manager to the unconscious, our self-aware consciousness took its recommendations from a salience network that switched, implemented, or directed whatever action or response was predicted to be most beneficial. And this is where the significance lies. Because by developing consciousness as an integrated but distinct counterpart to our unconscious mind, we had, for the first time, a choice. With this integrated unconscious-conscious mind architecture, we, perhaps uniquely, became able to deliberately make use of our unconscious, to choose when, why, and how to employ it. By calming the mind, reducing the noise of our inner world, we became able to harness insight and associative understanding that comes from a clear and "empty" mind and bring it back to the conscious brain in a way that allowed us to learn and elevate ourselves incrementally along Plato's spiral. In becoming conscious, we built a regulator, and through the use of that regulator, we brought into existence the single most important feedback mechanism we might ever encounter.

Out Here in the Perimeter

Our journey into the beyond is not yet quite complete. Life evolved in unconscious beings, organisms surviving through reaction to stimuli. Then came consciousness, in a range of forms, and our cognitive world became richer. Through intoxicants and mind-altering activities, we discovered ways to bring more of the unconscious into the conscious, but opening this door flooded our mind with the noise of sensory experience that was normally kept suppressed. That state is an informative state, but not a sustainable one. But with focus and effort we can calm the noise of the mind and, by keeping that door to the unconscious ajar, we can deliberately and purposefully shift our chosen reality to one that draws at least some of its substance from the realm of insight and understanding. The more deeply we can lean into that world, the more closely we might approach the state of pure awareness, and from there we might perceive *suññatā*, the emptiness of the void, in which all phenomena are without substance. In this realm, matter is no longer matter, but matter is energy, moving at the speed of light.

The universe as observable from Earth stretches roughly fifty billion light-years away from us, though we are not at the true center. Averaging all mass across this vast space, there are but five hydrogen atoms per cubic meter. Just five. The universe is inconceivably large and even more inconceivably empty. Yet the total energy in this void amounts to 10^{70} joules.[*] The energy and matter sparsely distributed across immeasurable space-time form planets that orbit stars, which in turn are organized into galaxies, then galaxy groups, galaxy clusters, superclusters, then into sheets, walls, and filaments separated by immense voids that, if it were possible to zoom out even further, would lend the universe a foamlike structure. Welcome to the cosmic web.

As far as we know, nothing can travel through this infinitude faster than the speed of light, nearly three hundred million meters per second. Staggeringly fast though this is, it is still too slow to allow energy from the very outer limits of the cosmos to have yet reached us.[†] As we look out ever further into the darkness, we look ever further back in time. The "reality" we define as our picture of the heavens is a composite of the past, conflated to our present. Earlier we saw how even such a robust phenomenon as gravity is far from absolute; now it seems that time also has little meaning. Since the groundbreaking discoveries of Einstein, we have come to realize that we live in a space-time continuum in which what we observe and measure is a function of where we are, what we are observing, and how we measure it. All forms of energy are affected by gravity, not just mass as Newton had proposed. So light bends, and time and space exist not as rigidly linear dimensions, but as flexible axes on a multidimensional graph. In chapter 1 we saw how the Greeks discovered magnetism and explored the ways that magnetic fields could influence other bodies. In the void of the cosmos, magnetism and gravity are the only forces that can reach across enormous distances to shape the structure of the frighteningly diffuse energy that lies out there, beyond. At the largest scales, magnetic fields permeate galaxies and galaxy clusters, their field lines arcing through space. And at the smallest scales, magnetic fields pull charged particles like electrons, distorting their circular orbits into spirals that become radio waves that we can record. It is thought that magnetic fields are a vestige of the creation of the universe,[53] spread throughout space in ways that controlled how galaxies formed.[54]

We inhabit the Laniakea galaxy supercluster, a region of space defined by clusters and filaments of galaxies whose overall motion would be inward

---

[*] 1×10^{70} is a 1 followed by 70 zeros. To give a sense of how big a number this is, if you counted the number of seconds since the birth of the universe, the "big bang," it would only come to about 4.3×10^{17}, just under half a quintillion.

[†] In late 2022, the James Webb Space Telescope captured images of galaxies dating to around 13.4 billion years ago—just 400 million years after the Big Bang (https://phys.org/news/2022-12-nasa-webb-milestone-quest-distant.html).

if it weren't for the inexorable outward pull of the expanding universe.[55] Galaxies spin, clusters flow, even filaments composed of galaxies—tendrils hundreds of millions of light years across—seem to rotate.[56] Elsewhere in the cosmos exist "strings"—tubular cracks in the fabric of space-time filled with nothing but pure energy—stretching across the entirety of the observable universe. In some places there are even strands composed of cosmic ray electrons, each strand 150 light-years long, regularly spaced like strings on an enormous ethereal harp. Astrophysics tells us that energy abounds, is everywhere, and is organized and structured, moving and flowing, permeating the void in electromagnetic wavefields whose properties we still barely comprehend. And here on Earth we are embedded within it, enmeshed in the colossal cosmic forest whose filaments reach and connect in dimensions more expansive than the ones we can readily perceive. Occasionally this energy manifests on Earth in ways that rock our physical, observable world. Dying stars send subatomic particles like muons streaming out into the ether, impacting planets like ours and irradiating life at the surface and destroying the three-millimeter-thick layer of atmospheric ozone that protects us from ultraviolet radiation.[57] These supernova explosions most likely led to widespread DNA damage in megafauna and direct mortality in the photosynthesizing microorganisms that formed the base of the food chain, collectively triggering abrupt periods of accelerated species turnover throughout Earth's history.[58]

What role could such destruction play? We might pass off such phenomena as misfortune, an inevitable but unforeseeable consequence of our chaotic passage through the energetic ashes of the life cycle of distant stars. But such a view sits uncomfortably with the apparent order, regularity, and structure of the vast fields of energy that we know we are a part of. The Austrian physicist Erwin Schrödinger, winner of the 1933 Nobel Prize in Physics, held that individual beings are simply aspects of a unified, higher agent and that what we experience as "reality" is, in fact, just a simplification of a shared consciousness. In this we see striking coherence with insights we saw earlier gleaned from psychedelic or meditative awareness; a common understanding reached through distinctly different ways of knowing. And Schrödinger was not alone in such a view. Another Nobel Prize winner, physicist Roger Penrose, published extensively on the idea that the physical laws of the universe are configured in such a way as to favor consciousness and, were the universe to undergo some form of cyclic renewal, that the laws governing each new instantiation of space-time could progressively evolve to "optimize consciousness."[59] Leaning heavily on the idea that quantum uncertainty could underpin the origins of consciousness in living things, Penrose and his coauthor Stuart Hameroff suggested that "there is a connection between the brain's bio-molecular processes and the basic structure of the universe," drawing direct analogies between their quantum physics–based conceptualization of moments of consciousness and those reported by Buddhist meditators. In both instances, consciousness is perceived as a flickering stream of "moments" that average forty per second, synchronized across neurons in the frontal and parietal cortex. The direct interweaving of quantitative insights from traditional Eastern ways of knowing with those from EEG correlates of consciousness emerging from Western medical ways of knowing is, at the very least, intriguing. Hameroff and Penrose go on to propose three possibilities regarding "the origin and place of consciousness in the universe." One option is the traditional Western science ("materialism") view, that consciousness in the universe has no distinct role to play. At the other end of the spectrum is the idea that consciousness in the universe is something outside of science, belonging only to notions of spirituality. But their third and favored interpretation is one that is scientifically grounded, but that also accepts the possibility that consciousness in the universe exists as "an essential ingredient of physical laws not yet fully understood."

In recent years, other leading cosmologists have picked up on this idea, proposing that our universe learns its own laws by exploring a landscape of possibilities, trying out permutations to discover what works best.[60] Like the evolution of life on Earth that we saw in chapter 2, evolution of the universe most likely encountered dead ends and periodic turnover of its constituent elements, but it also included a mechanism that allowed it to lock in, or accumulate, the changes that were found to be beneficial. Like Capra's cognitive cycle of adaptation and optimization, the unsupervised universe learned by altering its internal processes and actions in ways that allowed the available drive of energy to better flow through it. But it did more than this, because as we have seen repeatedly throughout this book, a successful learning machine needs the power of anticipation; it needs to predict. The autodidactic (self-educating) universe is one that integrates the present moment with stored information from the past to know what it needs to do next. And that definition is exactly the one we used earlier to explain our own human-based form of consciousness.

The entangled mycorrhizal web beneath the healthy forest floor, the coupled, synchronized neurons in the minds of conscious beings, and the filaments and clusters of electromagnetic energy that collectively shape and define the cosmos in which we live are each adapted, refined, and progressively optimized by feedbacks that switch focus, effort, or resources to wherever they are needed not just for the present moment, but for whatever's predicted to come next. These self-similar patterns of organization spanning atomic to galactic scales manifest in diverse and seemingly incomparable ways, and yet beneath the visible appearances of their physical forms lie invisible templates and processes that guide their operation in ways that, though hidden from view, survive and persist because of one common thing. Information. The mother trees share their wisdom with saplings in the form of electrochemical signals, bits and bytes whose messaging capability is comparable to the neurotransmitters in our brains, each axon in our brain pulsing with chemically induced electrical currents to trigger others to do the same, cascading a flow of information through one hundred trillion nerve junctions. But what form does this information take when we shift our gaze heavenward, out into the cosmos and into the realm of "pure energy"? Some researchers suggest that information is simply another fundamental quantity, like matter or energy,[61] and that if we can treat it this way, we can see equivalences between these forms. It has been proposed that a single bit of information has a quantifiable mass, around 3.19×10^{-38} kilograms, and that by acquiring this mass it no longer requires an input of energy to ensure that it persists—information becomes conserved. Maybe the dark matter of the universe is all information, and perhaps this information underpins the self-learning of the conscious cosmos, a process of continual creativity, something literally "in formation." Because to explore new possibilities, our present universe, as well as those that may have come before or will follow after, needs to retain knowledge of what went before. A consciousness that starts anew, unaided and uninformed, will assemble complexity far more slowly and with more dead ends than a consciousness that begins its assembly with structures once shown to have worth. By reassembling informational building blocks that themselves consist of smaller blocks and those of yet smaller blocks, learning is accelerated.[62] When innovating, choosing to copy past successes and to build from them proves far more effective than the continual search for innovation and entirely novel insights.[63]

The filaments of our conscious web extend outward to others as far as we choose to allow, and we might weave this silken tapestry as densely as we might like, embroidering layer upon layer in a multidimensional blanket whose threads ripple with the waves of our consensus reality. This is a visible world, a domain of knowing.

Tug a single loose thread and we might unravel our tapestry a little; pull too hard or too widely and we might distort much more of our blanket of existence. As with our cloth, so too with the intricate network of our forest, the electrochemically maintained symbiotic linkages each so vital to life yet so easy to break. We rely on redundancy, backup connections in case others fail. And we value species complexity, intergenerational sharing, and the nurturing wisdom that imbues our coupled community. This is a hidden world, a domain of inquiry.

Our conscious mind stretches to the heavens like the limbs of trees clawing their way upward to the light above the overstory. And still further above, our terrestrial web is mirrored in the firmament, composed of structures enlarging from galactic superclusters to walls and filaments, to the cosmic web of all matter, and outward to the furthest, unreachable, galaxies. Beyond this lies only the cosmic background radiation, pure energy in the otherwise "void and cloudless sky" of space-time. This is a barely visible world, a domain of not-knowing.

And yet all three are one reality, separated only in scale and proximity. And this unity is complemented with its twin, the inner world that is the world of the unconscious mind. We are unconscious beings having a conscious experience. Enhancing our reality by consciously engaging the quieted unconscious mind is the most powerful personal feedback we could ever wish to discover, elevating our existence beyond the sensory stimuli of the local here and now to one that is unified and integrated with all life and with a universe that learns. The spirit that we discover there is one that connects us to one another in what we might call the human spirit; the wisdom we find there is one that we possess innately, coalescing from the totality of sensory stimuli and cognitive associations that we grapple with continuously without knowing, a reality more complete than our conscious mind can perceive. This is the domain of understanding.

Mostly we live in a consciousness whose reality is narrow, focused, and weighted to guide us in ways that enable survival. But through art, music, and guided psychedelic experiences, we can glimpse fragments of the overwhelming sensory orgy of Plato's "true beauty" of reality, and through meditation, fasting, dance, sex, or other exhausting feats of physical endurance, we can adapt and orient ourselves in ways that suppress our noisy conscious mind. Through conscious choice we engage the unconscious and allow the wisdom of the infinite to reveal itself to us so that we might, fleetingly, taste the eternal delight of enlightenment.

SYNTHESIS 8

The universe is full of magical things patiently waiting for our wits to grow sharper.

—EDEN PHILLPOTTS, 1919

In 2015, three scientists published a paper in *Physical Review E*, one of the most highly regarded scientific journals in the field of statistical physics, which presented the results of simple but ingenious experiments involving some ball bearings and an electrical current. Chrome beads were placed in a petri dish with a small amount of viscous oil. Above the dish was a powerful electrode capable of transmitting up to 30,000 volts, with another electrode (a closed metal loop) placed around the beads within the dish. During the experiment the electrical current "jumped" from the upper electrode to the one below, but in doing so made use of the metal beads in the dish, just as lightning preferentially makes use of any available antenna. The flow of energy through the beads caused them to move about in the oil, each movement slightly affecting the electric field and changing the motions of each of the other equally entranced beads. Curiously, what emerged from this excited jostling was something even the researchers themselves didn't expect. For the beads didn't simply dance around the oily dish at random, cavorting in an energetic whirl of unpredictability. Instead they organized themselves into a single long line that connected at right angles to the ring electrode in the dish. Nearer the center of the dish, beneath the source electrode, the beads branched from this trunk so that the entire arrangement formed a treelike structure. And this tree moved, feeling its way around the ring, swaying from side to side as it did so. If broken by the experimenters, the tree reformed—it healed itself. From the variety of behaviors they observed, the group were forced to conclude that this simple system of electrically conducting metal beads exhibited "properties that are analogous to those we observe in living organisms."[1] And the reason for this magical phenomenon? Efficiency. The system had arranged itself in a way that allowed an efficient flow of energy through its structure and then continued to adapt and rearrange itself in order to seek out better configurations that could even more effectively harness that flow.

If inorganic systems can self-assemble to optimize the flow and utilization of available energy, what might this tell us about the origins and subsequent evolution of biological systems—the collected cohort of organic phenomena that we call "life"? Recently, a new approach to understanding biological life from the perspective of fundamental physical laws has been put forward, one that offers a way to contextualize the evolutionary process that differs from the traditionally adopted Darwinian theory of natural selection. Jeremy England, a physicist and the head of artificial intelligence at GlaxoSmithKline, published a theory that puts the emergence and evolution of life in a new physics-based framework. At the heart of his theory is the concept of dissipative adaptation. In dissipative adaptation, random populations of molecules, when exposed to a flow of energy (a "drive," as we have seen previously), tend to self-organize in a way that allows them to more efficiently absorb that energy and redirect or dissipate it. Using tools from statistical physics, wherein the collective properties of individuals within a population tell us about the functioning of the system as a whole, England proposed that organization, indeed life itself, might emerge from the seemingly chaotic collisions and reactions of individual molecules that take place when they are exposed to a flow of energy. The key thing is that the system adapts in ways that better align with the drive, in turn changing the way that other parts of the system respond, each time such events take place. Gradually, this process of adaptation optimizes the system as a whole for the environmental conditions it is exposed to. From this incremental process, England notes, "life manages to squeeze exquisite reliability in behavior on large scales from a jittery herd of individual molecules."[2] Most importantly, each successive structural adaptation is locked in so that the system as a whole progressively evolves. From the outside, this incremental evolutionary path appears deliberately engineered, intentionally shaped, or guided by an "invisible hand."

The feedbacks that shape these emergent phenomena of self-organization and optimization are processes that rely on an external drive of energy; they embody it and make it their own. By combining adaptation with irreversibility, these feedbacks allow a structure or a system to evolve. But that evolution is not one of blind and random change. It is directed by the energetic flows that sustain the system as a whole. Like iron filings compelled to align with the field lines of a magnet, evolving systems across a wide range of complexities are similarly compelled to restructure themselves in ways that are dictated by forces outside, and within, themselves. Wood-rotting fungi such as *Phanerochaete velutina*, for example, send out threads that forage for nutrients.[3] Those hyphae that discover food are strengthened by positive feedbacks, reinforcing that connection. But the filaments whose foraging is unsuccessful are pruned back and die off so that over time it appears that the fungus has migrated only in the direction of the food, as if it somehow knew in advance. But this is not so much a signature of divine guidance as one of reinforcing feedbacks. Similarly, the defensive action of honeybees and the swarming of foraging ants also rely on reinforcing feedbacks to rapidly achieve the desired goal. Just like the foraging fungus, ants radiate out prospectively in search of food. And as they do so, they secrete hormones. Those ants that find food then return to their origin, following their own scent trail, all the while still secreting more of the same hormones. By the time they have returned to the colony, their hormonal trail is now twice as concentrated as that of other trails whose creators have yet to return. Other ants, on leaving the colony, are now attracted to the stronger signal and follow it, each new insect adding its own secretions to the route. A regime shift rapidly ensues, in which one path above all others rapidly becomes the preferred route for the colony as a whole. With no predetermined vision, the feedbacks embedded in their foraging process allow ants to efficiently explore and exploit their landscape of possibilities. Inside the human body, swarms of white blood cells called neutrophils migrate to sites of tissue damage or infection. Like ants, migrating neutrophils release attractants that encourage others to follow, each time amplifying the signal and leading to an exponential increase in chemical concentration at the site of interest.[4] But eventually there are enough neutrophils at the desired site, and they need to stop migrating, stay where they are, and do the job they were employed to do. Switching off this swarming behavior now requires a negative feedback. The swarming cells become increasingly desensitized to the high concentrations of the attractant chemicals, remain where they are, and no longer attract others to join them.

Finding Feedbacks

The swarming behavior of ball bearings or colonies of insects relies on the coordinated activity of a group of individuals who, through the action of feedbacks, self-organize for a shared purpose. And the motivation for that self-organization, the drive for the necessary feedbacks, comes from an external flow of energy. In this book we have explored a diverse suite of phenomena that span timescales from milliseconds to billions of years and physical dimensions that stretch from the subatomic to the cosmological. And yet in each of these vastly different realms we have discovered the same kinds of processes, each time manifesting in its own unique way, but each also sharing its behavioral traits with many others. Let us now, then, revisit those earlier investigations and see if we might draw from them a suite of common signatures, threads of a fabric that binds and connects all of life and the physical space we inhabit.

In *Earth*, we descended like Dante into the infernal depths of Hades and Tarturus, into the "funnel-shaped pit" that led us to the inner levels of our home. In the Earth's mantle, that thick layer of molten rock, we found giant tree-like plume structures that brought superheated basaltic rock to the surface. As it cooled and cracked, this rocky skin fractured, self-organizing into an arrangement of plates that most efficiently allowed the hardened outer shell to accommodate the hellish forces beneath. From the gravitational forces acting on that network of plates, coupled with the flow of energy upwelling from Earth's core, arose the tectonic movements that slowly but steadily began to move, subduct, and recycle our lithic shell. This tectonic drive gave rise to earthquakes and volcanoes, a coupled system whose detailed operation still holds secrets, but that we know was accelerated by the lubricating feedback of sediment, water, and ice. By up-lifting land, tectonic processes exposed rocks to the early atmosphere and allowed chemical reactions to take place that drew carbon dioxide out of the air and locked it away in the deep ocean. During its hottest times, rain-fall events on the early Earth followed a sawtooth pattern as moisture in the air built up gradually until it could no longer be contained, releasing back to Earth in torrential deluges. This oscillator allowed the imposed drive of energy to incrementally accumulate until a critical state was reached. Lightning during those times was more frequent than it is today, and in striking the barren earth, those electrical storms created phosphorus, which played a key role in the origins and establishment of life.

And once life became established, first in the oceans and then later on land, it too exerted a feedback on the atmosphere. Photosynthetic algae captured carbon dioxide and released oxygen, cooling the climate and producing pulses of oxygenation that gradually extinguished anaerobic life and paved the way for an explosion of complex life 541 million years ago. And throughout the 4.53 billion years of Earth's existence, cyclic patterns of change pushed our continents, oceans, and climate in ways that allowed for repetition with adaptation, cognitive feedback loops that incrementally nudged our planet into a more stable state that was increasingly optimized for the emergence, proliferation, and evolution of intelligent and sentient life. As our physical Earth evolved, the ecological niches available to the rapidly diversifying biology on its surface also changed. The balance between speciation and extinction waxed and waned over tens of millions of years, occasionally bursting forth with frivolous experimentation but at other times abruptly curtailed by catastrophes such as volcanic eruptions or meteorite impacts. Life evolved under a constant flow of external energy, in which episodic mutation allowed the most efficient traits to be locked in and retained, while those less well-aligned with that flow died out and were lost. Assembling multiple components from prior adaptations that were successful, life built complexity in new organisms, all the while continually absorbing energy from the environment and responding to that external drive in ways that allowed each new life-form to be more efficient. Through cooperation and mutual dependency and the self-limiting feedbacks of predation and resource exploitation, our complex dynamical system of biological life incrementally self-organized to a state in which the distribution of biomass within discrete size ranges was, by the time modern humans appeared, perfectly arranged across twenty-five orders of magnitude.

The breath of Gaia that brought life to our Earth gave us our *Climate*—not just an atmosphere, but a process, a means of living, a constant recirculation of the essential force that flows through all things. This spirit, the pure consciousness of Atman, evolved and adapted as the physical and biological systems beneath it jostled and hustled, conniving and contriving to oppose and dominate one another. In the growth of lush forests three hundred million years ago, our biosphere locked away carbon dioxide that had previously kept our climate much hotter than today and in doing so changed not just the atmosphere, but also the biodiversity that depended upon it. As the violence of the early Earth dwindled and life at its surface evolved into one of balance and harmony, fluctuations in solar energy arising from our planetary waltz around the sun gave rise to patterns of cyclic climate variability that set the tempo for regular ice ages. Through the regularity of oscillating global temperatures arose repeated opportunity for adaptation and optimization, and because components of the Earth system retained sufficient "memory" to prevent the governing thermostat from producing a quivering state of frustrated anxiety, Earth's climate optimized progressively toward the sawtooth pattern of the relaxation oscillator we are governed by today. But the time-delayed feedback of the carbon dioxide once sequestered in Carboniferous swamps and now released back to our atmosphere through the burning of coal, oil, and gas now threatens to exert a drive of its own. And this new drive will be sufficiently powerful to dictate the adaptations and optimizations that characterize the future evolution not just of our species, but of all others as well.

We *Humans* are the result of more than four-and-a-half billion years of terrestrial, biological, and climatological organization, adaptation, and optimization. But it was only seven million years ago that our ancestors moved from the trees to the savannah plains, when the flow of solar energy changed the landscape in ways that encouraged new ways of living. As we used our walking skills more frequently, we became better able to exploit new resources, which established a reinforcing feedback that encouraged us to stand progressively more upright. We roamed further and adapted our lifestyle to seek out energy-dense foods like meat. To find this new food, we organized ourselves into collaborative hunting groups and developed ways of communicating, planning, and predicting. Through tracking we established art, then science, and through the growing interaction with one another we defined languages—systems of communicating that locked in associations between sounds and meanings. Neural synchronization, facilitated by speech, strengthened mother-child bonds and brought solidity to our familial groups. The tools and skills we developed we also began to share with others beyond our group, and the feedbacks that arose from these social interactions steadily shaped our evolution. As the external drive of climatic variability continued to sometimes push us to migrate into new territories, it in turn allowed our populations to mix and demanded that we establish shared beliefs and values. Social contracts and the diffusion of ideas led to a phase change in our cultural evolution that, from initially slow beginnings, rapidly accelerated the pace of our individual and collective knowledge. We sang the world into existence, and in doing so awakened our inner selves. Through cognitive and cultural developments, and in our connection to the Earth through a shared soil and gut microbiome, we progressively optimized our physiology and behavior to make best use of the environment and the drive of energy that perfused it.

As our global population expanded, we began to rely less on nomadic subsistence, and found advantages instead in fixed settlements and a close *Society*. Establishing a place for ourselves, we built clusters of houses and buried our dead in stone tombs, dug mighty earthworks and adorned them with lithic monuments that helped us fix our place not just in space, but also in time. Devising calendars to track the annual movements of the sun, we gained predictability for the farming practices we were beginning to adopt, but the colossal efforts involved served another purpose beyond the functional keeping of time. Our shared labors brought unity and cohesion to our societies, immense efforts illuminated and guided by the light of a single vision. But despite our endeavors, we had still to contend with the unpredictable—the wrath of Zeus meted out through natural disasters or climatic change. How we were impacted by these events depended on the feedbacks that cascaded through our societies, shaped in part by the types of personal connections we had, the diversity of the crops we grew, and the preferences and policies of those we had entrusted with positions of power. Being able to absorb or dissipate the energy flowing through our societies, whether sourced from outside or from within, was what enabled us to build resilience. Yet that resilience could be tested by wars, persecution, disease, or famine, and the memories of those events shaped how we rebuilt our lives and those of the generations that followed. Through synchronized behaviors, entraining our actions and views with those of others, we developed empathy, compassion, and altruism. By balancing the opposing forces of desire and need, we were able to find stability in a landscape of social change, and together we evolved a collective intelligence that fostered learning and purposeful growth.

Secured in a society built on the progressive adaptation and optimization of billions of years of evolution, we are now sentient, conscious, and aware of our *Self*. But this single self is a composite of visible and invisible, physical and nonphysical worlds. From the flow of external energy that we perceive and absorb, we gain experience. And through experience we find knowledge. But it is only through the illumination of the nonphysical world, the energy that flows within us, that we might find a deeper understanding of a wider, yet hidden reality. Our bodies thrive on the flow of energy that we direct toward them, an energy that fuels the regular and dependable beating of our hearts. Groups of pacemaker cells, repeatedly charging to a critical state, discharge en masse in a synchronized pulse of contraction that ensures viability and rejuvenation for every one of our forty trillion bodily cells. Our vagus nerve collates signals from our body and transmits them to our brain, where the effectiveness of our sensory processing as well as the state of our mental health depend critically on avalanches of synchronized neural activity. But feedbacks sometimes work against us, earnestly reinforcing connections within single brain regions that should normally be more widely connected to other areas, leading to emotional problems that trap us in suboptimal behaviors. For effective mental functioning, we rely on brain synchronization across widely separated regions and communication between those regions that self-organizes to a state close to the edge of chaos, where information transmission is optimized. By integrating brain activity and bodily sensations, we absorb a diverse stream of incoming information, contextualize it with memories of past experience, and use that synthesis to guide our behavior in ways that seek to fulfill the predictions we make about the world.

But what lies *Beyond* that physical self, the visible body in which we reside? We perceive our existence as one lived individually, but we are connected electromagnetically to every other living thing. It was this invisible connectedness that allowed culture to arise in early human societies and perhaps even underpins the origins of consciousness itself, yet these days we seem unable or unwilling to feel it and understand it. We advance through associative learning, not just individually but collectively, through the push and pull of prediction feedbacks that continually enable us to update our perception of reality. Our consciousness exists along a continuum, and the reality we inhabit at any given time is the one that we choose to inhabit through the self-reinforcing feedbacks associated with the choices we make. And we can choose to alter our reality, to widen the field of sensory experiences in ways that lessen our attachment to a sense of self and reveal to us a more tangibly felt connection to the wider sphere of life. Through feelings of awe and experiences that transcend our everyday experiences, we feel a pull toward something bigger than ourselves, a sense of a liminal spirit that our brains are hard-wired not just to acknowledge, but also to use. Because by consciously quieting the unconscious mind, we can engage the most powerful personal feedback of all—one that can help guide us and help us learn through understanding, not just through knowing. And through this understanding we find connection to a universal field that pervades the cosmic web, a stream of pure energy that, perhaps, is composed of information that directs a self-learning universe, continually optimizing itself for the emergence of ever more complex life.

The diverse manifestations of that suite of feedbacks all share commonalities, both in how they arise and how they work. Above all, they need a drive, a flow of energy that pushes a system away from the equilibrium state it would find itself in if no external forces were acting upon it. Pushed in this way, many of the systems we have seen tend to fluctuate through cycles of change, rising then falling, repeating broadly the same path each time, but each time passing through states that differ sufficiently from one another that new opportunities are continually presented. Often these cycles of change are asymmetric, they rely on a longer period of buildup than collapse, tracing out a sawtooth pattern or an asymmetric oscillator. These patterns of change allow components of complex systems to gradually become entrained with one another, to align their cyclic fluctuations in ways that eventually regulate individual changes to be synchronized in time. And as this synchronization diffuses throughout a system, it organizes that system in such a way that, to an external observer, it appears to have been intentionally orchestrated. Through the push-pull feedbacks that bring about synchronization in time and self-organization in space, systems adapt their functioning to make best use of the available drive that fuels them. By locking in beneficial changes and through cyclic repetition to test and refine those changes, the process of incremental adaptation leads to optimization, to a system that is as efficient as it can possibly be, yet one that also allows for growth.

This depiction of how the world works brings together theories of self-organization,[5] the emerging science of synchronization,[6] the "cognitive cycle" described and explored by Fritjof Capra,[7] and Jeremy England's theory of dissipative adaptation.[8] Each of the physicists and mathematicians behind these advances has explored the world in rigorously quantitative ways and yet has found a beauty that transcends the sterility of numeric analysis. A beauty that reflects something larger than science, an ineffable something that has been the muse of artists and philosophers for millennia. In synchronization, we might find, for example, an analogue in rhythm—movement or sound coordinated in time. In self-organization, we might recognize the concept of harmony, each member of the scattered orchestra coordinated yet physically apart. And through the unity of these two phenomena, our system continually makes, tests, and updates predictions as to the adjustments that are needed to optimize its overall functioning, an optimization that strives for the essential beauty to which Plato alludes, the divine quality that radiates through nature in all its forms. And it is this triad—rhythm, harmony, and beauty—that artists like Harold Speed suggest might allow us to develop a relationship with whatever it is that is beyond us, the fundamental essence that "is felt to be behind and through all things."[9] Nobel Prize–winning physicists such as Erwin Schrödinger and Roger Penrose, peering into the emptiness of the cosmos, interpreted the traces of that primordial force as evidence of a quantum wave field, an all-pervading source of energy that exists as a potential, transformed into an observable state of being only when that wave field collapses back to a measurable point in space and time. In Eastern ways of knowing such as yoga, this field becomes prana, the divine life force, an inward-flowing energy that is the "drive" for both our physical and nonphysical self. And just like the quantum wave field, this flow of energy exists as a potential, something that is there but something that needs to be welcomed and cultivated in order that it might be used most powerfully.

Finding Purpose in Life's Rich Pageant

How, then, might we make use of these insights, the things we have learned from the distillation of observations from plate tectonic and evolutionary processes, from collective and individual behaviors, from neurological and cardiac activity, and from the spiritual and cosmological forces that invisibly shape and guide our lives? The feedbacks we have seen either accelerate or suppress a process, and in doing so alter the interactions among parts of a system. And feedbacks can provide the opportunity to start over if the direction of adaptation is less than optimal. Without feedback, there's no change, no betterment. And without forcing, without a drive, there is stasis. A sustained flow of energy pushes a system to adapt and modify itself to better suit its external or internal environment, an incremental response that reflects learning or cognition. The feedbacks that take place in the driven, highly nonlinear, system that is our everyday life, therefore offer us opportunities to learn.

As we go through the innumerable learning loops of life, some small, others more significant, our cognitive cycle of adaptation and refinement relies on harnessing an external flow of energy via sensory stimuli. But we must integrate those sensations with memories of past events if we want to make predictions that can help us adapt and change. Memory, then, is essential, but it is not the fixed and absolute thing we might think it is. Our memories can change depending on how we recover them,[10] and through consciously exploiting this unconscious resource, we can learn to find something positive even in negative past events. We can progressively reframe our past into whatever we want it to be, and if we are willing to invest the effort, we can establish long-lasting changes in our perception of the past and find positive outcomes that in turn change our predictions for the future. When we see the potential rewards for the mental effort involved in such relearning, we come to intrinsically value that effort, to recognize that what is most expedient is not always what might be most beneficial.[11] Optimizing our self or our society is, therefore, not always about doing what is easy, but doing what will most help us in the long run. And this is why our Neolithic ancestors built stone circles, and why President John F. Kennedy chose to "go to the Moon in this decade and do the other things, not because they are easy, but because they are hard."[12]

Choosing what is hard is a temporary sacrifice for long-term gain, an investment we intentionally make so that we might advance our situation in some way. Angela Maxwell knows what it is like to choose a path that is hard. In May 2014, she left behind her previously normal life and started walking. Over the next six-and-a-half years, Maxwell walked twenty thousand miles around the world, a journey of self-discovery and adventure that was motivated by a desire to find a "deeper connection with nature and people."* And she isn't the only one to embark upon such feats—at least ten people are known to have circumnavigated the globe on foot, with many more undertaking similarly arduous endeavors. All of these undertakings, like many of our everyday decisions, require a commitment to a certain action and to the planning and execution of that action, driven by a desire to accomplish or discover something, to connect with a deeper purpose. The drive for such quests, their planning (a form of prediction), and their execution are all elements of the learning loop, the cognitive cycle that is now familiar. Through investing in what is hard, we reveal what was otherwise hidden. We commit to a challenge, with faith, in order that we might gain a more complete understanding of ourselves. We strive not just to exist, but to develop.

Development, cognition, or learning—however we choose to frame it—is often seen as something that takes place within our neural architecture, within a brain whose structure and functioning allow the growth of our intellect. But cognition and learning are phenomena that transcend scales and are not just bound to the bodily hardware we most often associate them with. Sometimes, cognition and learning arise in ways that may not at first be recognizable as signs of intelligence. In what has since become known as the Santiago theory of cognition, Humberto Maturana and Francisco Varela argued that all living systems are cognitive systems, that living is a process of cognition because it is through adapting to its environment that an organism learns how to survive.[13] A key aspect of their theory is the idea that, in any system whose components are interconnected, feedbacks between those components enable the system to adapt to external forces in ways that give it agency—the potential to shape its own future. But to ensure survival not just of itself, but of its species, an organism not only needs to extract energy from an incoming drive and use that to encode information in its structure, but it must also propagate that

---

* www.bbc.com/travel/article/20210527-the-woman-who-walked-around-the-world

information forward in time.[14] Given the extra energy needed to encode and decode descriptive information, organisms look for a shortcut; they seek efficiency. Like the compression algorithms we use in our brains and in our computers, living systems attempt to minimize the amount of information that they handle by simplifying what they need to process. And it is these shortcuts, these data storage optimizations, that give rise to the symmetry and beautiful simplicity of nature.[15] By focusing on sequences of repeating similar patterns, rather than more complex arrangements with unrepeating variability, life finds a way to harness the driving force it is exposed to and pay it forward in the most efficient way possible.

And if repeating discrete patterns, in the form of chunks of encoded information, works well for individual reproduction, it is easy to see how the optimization of a species, or even the evolution of entirely new forms, might efficiently arise from the addition, subtraction, or recombination of such informational chunks. Chapter 7 touched on the concept of assembly theory, a new way of thinking about the origins of life and subsequent evolutionary development. One problem that has always dogged the idea of Darwinian natural selection is the inconceivably vast number of chance mutations that are possible and the low probability of highly complex life-forms emerging from that infinity of possible mutations within the geologically short time frame available. The solution to this apparent problem requires us, once again, to invoke the concept of memory. Through memory we devise a way in which prior success can be identified and used to inform future designs. If we imagine the evolving complexity of life as a process in which previously assembled chunks of information—biological building blocks—are simply reused in different combinations or sequentially aggregated into new compositions, it seems logical that the designs that will be reused or copied most will be those that are most suited to their environment.[16] Even if a particular organism goes extinct, the building blocks from which it was constructed may continue to exist in other species and be replicated or recombined in subsequent species yet to evolve.

A drive, modified by feedbacks, allows cognition. And through the integration of memory into that process, each successive learning loop, each adaptive feedback gives rise to a new and emergent phenomenon—something we call *meaning*. Through assembly processes, life distills meaning from initially random experimentation. Searching the genetic landscape, assembly builds ever more complex life from the most widely copied fragments of information and enables advances to occur far more rapidly than blind or random change would permit. And as with our physical development, so too with that of our intellect. Because life not only searches the genetic landscape to find meaning in the most preferable of the options available, but it also seeks out the cognitive abilities, experiences, and skills that offer to a population the advantages that might best ensure its advancement.[17] Switching from searching new areas (exploration) to searching within an area (exploitation), this collective cognitive search process naturally tends toward goals with shared meaning, advancements that favor the group as a whole, not just the individual. The wider the field for either exploration or exploitation, the more numerous the options for novel group-level enhancements. Viewed in this way, the ever-increasing population densities we live among and the inexorable rise in global connectivity we enjoy most likely mean that the advent of an evolving global-scale "shared consciousness" is entirely conceivable. Because the energy that provides the external drive of our existence is the same energy that fuels all living things, and the consciousness that we experience exists on the same sliding scale of consciousness that those other species also experience. Life, in all its forms, is instantiated from probabilities in a unified quantum wave field of energy, and each instantiation, each individual, is driven to use that energy and to find meaning. When we talk about learning, it is not only ourselves who are seeking development and advancement, it is this wider consciousness of which we are only one small part. We are a component, not just of a planet-scale intelligence, but of a *planetary intelligence* whose search for meaning relies on cognitive feedback loops that are "global in scale, coordination, and operation."[18]

This *anima mundi*, this world soul, is the Buddha-nature of Mahayana Buddhism, and the Brahman Atman of Hindu beliefs, the inseparable universal and individual soul. And because prana, the life force that sustains all things, comes from this fundamental field, it completes our single cyclic system. Prana is what provides the drive for the emergence and maintenance of life on Earth, and as life adapts its structure to better absorb that drive, it experiments, reshapes, and evolves. In doing so, each living thing embarks upon a spiral of learning whose advances are metered only by the pace with which it can adapt and grow. Collectively those individuals raise the level of group knowledge and understanding, and through the aggregated efforts of learning emerges meaning and connectedness. In recognizing that connectedness, we discover the shared consciousness and the life force that pushes us slowly, unrelentingly onward. One more time around the sun. Once more to the dark side of the moon. One more beat of our heart. One more breath to fill our lungs.

The Significance of Information

So far, we have painted a portrait of life and of the universe in which meaning emerges incrementally from innumerable learning loops that are nested
within each other, their consequences cascading through systems from
particle to cosmological scales, driven and directed by embedded feedbacks
that balance, suppress, or accelerate change when needed. Within each
loop of cognition there is adaptation and change based on a prediction that
what worked well last time will work equally well next time. But life and
its systems can be chaotic, and emergent phenomena threaten to undermine
the security and reliability of our incremental growth. In defining our own
path in life, as we saw in chapter 7, we might choose to carefully regulate
how much we let ourselves be directed by purely external sensory stimuli,
by the badgering and relentless noise of what is "out there" in the chaotic
world. Because there are other stimuli from within that also request our attention, a calling from our physical body perhaps or from the thoughts that
form in our minds as we endeavor to navigate life's challenges. To maintain emotional stability and personal resilience, we might seek to balance
these differing demands, to follow a path that acknowledges them all, yet
lends unto each only as much import as necessary. To find this balance, the
meaning amid the noise, we can turn inward to what is "in there," in our
unconscious mind. Quieting the cacophony of the wider reality to which
our unconscious is privy, we find clarity in the calm, insight in the void,
and understanding in the emptiness. Reconnecting with ourselves, we
might then employ that clarity, insight, and understanding to engage more
purposefully with the visible and physical world, setting intentions that are
guided by something bigger than ourselves. We become agents in our own
cycles of learning, controlling the ever-present feedbacks in ways that help
us progress in whatever direction is most beneficial not just to ourselves,
but to the planetary intelligence of which we are part.

We optimize our condition based on the experience and understanding
we have, but as we go forward, we must also weather the storms of the unpredictable, finding resilience in our structure and in the connections that
sustain us. And in doing this, we still rely on the continued testing of our
predictions against reality, the perennial updating that is necessary to ensure
congruence of our reality with that of those we share our existence with.
We learn to rely on information, and we convince ourselves that "knowing more" will improve our situation. But in the complex systems that
typify life, it is not always possible to accurately predict outcomes based just
on information-guided theories of cause and effect, because uncertainties
in our evidence often mean that multiple theories could be equally plau-

sible.[19] In our rapidly advancing technological world, we now turn instead to information processing techniques that dispense with theory entirely and parse monstrous streams of data in ways that look instead for patterns and connections, clues as to how, why, where, and when information is transmitted and what those patterns might tell us about ourselves. By analyzing nearly a billion time-stamped events across a suite of social media platforms, researchers found that information seems to flow in bursts, or avalanches, just like the self-organizing behaviors we have seen previously in earthquake activity, in the pacemaker cells of the heart, and in the neuronal activity of our brains.[20] Without even needing an explanatory theory, data analytical techniques such as these can tell us that information transmission across social media platforms, the timing and frequency with which we communicate by email, or the way that we govern the schedules of our daily lives all operate just like many of the natural systems we've already explored, close to the edge of chaos.[21]

But as we confidently stride headlong into a new era of gene editing and cellular engineering, machine learning and computational intelligence, how sure are we that the next significant rotation of the universe's learning loop will be one that doesn't dispense with our (human) form of intelligence completely? What do we, as a species, still have to offer in a world where pattern-recognition software learns from data faster and more accurately than theory-guided scientists? Where a computer can now write code at least as well as half of a group of programmers competing against it?[*] And where artificial intelligence software can be trained to learn not just how to interpret data, but to do so in ways that are guided by human values?[22] If computational systems can develop beyond simply seeing patterns in vast and multidimensional datasets and begin to consider value and significance, can they, too, find meaning in what they are doing and impart some novel element of creativity into the outputs they produce?

Currently, perhaps reassuringly, it seems not. Although the technology is immensely impressive, the systems that, for example, have completed Beethoven's 10th Symphony[†] or won fine art competitions[‡] are still hampered by the training on which they depend and on the prompting and guidance of a human being. Learning from existing information means that whatever such tools produce, it is always derivative, tending toward

---

* www.science.org/content/article/ai-learns-write-computer-code-stunning-advance

† https://theconversation.com/how-a-team-of-musicologists-and-computer-scientists-completed-beethovens-unfinished-10th-symphony-168160

‡ www.vice.com/en/article/bvmvqm/an-ai-generated-artwork-won-first-place-at-a-state-fair-fine-arts-competition-and-artists-are-pissed

an average of the patterns used for their instruction. Humans may be losing primacy in certain areas of science, but it is the current inability of computers to produce something truly new that, for the moment at least, means we still have the upper hand when it comes to the arts.

Finding Beauty

The arts, then, are where we remain most human. And because of that, they have the greatest potential to build community. By enabling and strengthening a community spirit—the real-life interaction between people through the shared experience of exhibitions, concerts, or performances—the arts connect people physically and emotionally. Bringing together individuals at the same time, in the same place, and at the same event, the arts facilitate synchronized neural activity and suppression of the parietal lobe, increased alpha wave activity that fosters connection, empathy, and a diminished sense of ego. The sensory experiences with which we engage trigger the release of dopamine from our brain's reward circuit, instilling an individual feeling of pleasure that we experience in the company of others. We find unity in that shared experience, in the shared emotions. And each cycle of reward secretes dopamine both in anticipation of pleasure and at its peak, dropping rapidly to begin again like the sawtooth oscillator we've encountered so many times before. Through repetition we achieve progressive entrainment with those present in our proximal group, eventually enabling the totality of the collected audience to respond as one—a single, unified consciousness in which self-other boundaries are blurred and a sense of kinship and attachment emerges. And it is by bringing about this wider sense of connection that the arts can play such a vital role in our societies today, just as they have done since the Paleolithic. At a time when scientific messages such as the need for collective action to address climate change or biodiversity loss need not just to be heard, but to be taken up and acted upon, it is the arts that can most tangibly achieve this. Not by "translating" or communicating the scientific message, but by reminding us that we are more than just a collection of individuals. Our messages will be more readily heard by a group that is unified than by the same individuals isolated in their homes, engaging only in the arts through fragments posted on social media, each of us distracted by a barrage of sensory impulses that each compete for attention but that are each devoid of meaning.

Art brings meaning to our lives, because it "deals with ideas of a different mental texture, which words can only vaguely suggest."[23] In art we recognize beauty, one of Plato's highest ideals, one that incorporates knowledge of mathematics and the appreciation of concepts such as regularity and symmetry. "There is a strange and wonderful mathematical order in physical phenomena," wrote E. F. Schumacher,[24] one that manifests a "severe kind of beauty" and a "captivating elegance." But, Schumacher noted, "it has no warmth, none of life's messiness of growth and decay, hope and despair, joy and suffering." Like the artistic creations of artificial intelligence, the mathematical beauty of ideal forms is one that comes from a life led "according to concepts"[25] where art has "sunk to the level of pure entertainment." For art to have meaning, therefore, it must embody beauty in ways that convey something beyond regularity and mathematical predictability. So although the most enduring artistic styles are those that retain a grounding in physical and biological constraints, like balance, harmony, and perspective, they are more than just these things. Great art leaves space for the viewer to build their own interpretation, to bring to the piece their own understanding.[26] And it is in the process of meeting halfway, in the shadowy space between the artist's truth and the viewers' beliefs, that we find meaning and understanding.

And so we see here a glimpse of what it is that imparts significance to beauty and a sense of magic to great art. It is the *process* that we go through, not the end in itself. Nietzsche describes art as a consolation for the horror of life, a horror that he felt was the true reality. It is through art that we might find meaning in this "horrific" process of life, a form of self-knowledge that enables us to find understanding of ourselves and of our place in the world.[27] The *I Ching*, the Chinese Book of Changes, describes beauty as an expression of the way (Tao). Tao is not so much about achieving anything in particular, but about moving through life with ease and with grace, focusing on form, not content. Art that concerns itself with content at the expense of process, grace, and form is art that is impoverished and devoid of the beauty that brings meaning to our lives.

Too often we find ourselves living in an atomized society in which the arts have been reduced to just such fragments of content, wrested from their original context and pushed onto us in a chaos of sensory impressions, each one vying for attention in an already crowded arena of sensationalism, artifice, and hyperbole. In such arenas, we have reduced the arts to nothing more than a spectacle to behold, a moment's entertainment and nothing more. In doing this, we deny the arts their importance as ways of knowing and their relevance as a language that allows us to relate meaningfully to one another. Where science can give us knowledge, the arts could grant us pathways to understanding. Through the arts, we catch a glimpse of Plato's hidden beauty, the pure reality that we find at the boundary between our conscious and unconscious mind. Art can cleanse the doors of perception and transport us to other times or places, allowing us to transcend the noise of the eternally frantic external world, and guide us inward instead, on a contemplative journey of inner seeking and discovery. Both to Dante and to Plato before him, true art, and the beauty embodied within it, were *the way*; they were the intermediaries that engaged human faculties in a manner that brought a person ever closer to their higher potential and drew their soul toward philosophical contemplation and to the level of understanding needed for the perception and appreciation of absolute beauty. It is in the pursuit of just such a state that traditional ways of knowing recognized what it means for a thing to be "good." Good things were those that brought a person closer to the transcendental quality that extends beyond thinking, beyond emotion, into an unconditional love that persists in spite of condition, state, or circumstance. And it is here, in that wide-open state of consciousness, that we extend our notion of what is possible.[28] If "the world is a work of art"[29] and the "beauty of a work of art is in the work itself,"[30] then the transcendental beauty of our existence lies not in the content of our lives, but in the grace we bring to the process of living them.

We live as a cycle of change, embedded among other cycles that iterate through innumerable revolutions, evolving through learning, combining ideas and experiences and finding meaning amid the chaos of sensations and appearances. Our loop of learning is directed only by the understanding we acquire and the quest for wisdom we might pursue. But regardless of our destiny, our progress is fueled by an unshakeable drive, an all-permeating field of energy that flows through us and through every other thing. As we learn to adapt ourselves to that flow, we dissipate the charge more easily and allow it to serve its function more effectively. That drive is information, and in letting it pass through us most readily, we broaden its reach and accelerate its spread. In passing through us, each drive pushes us imperceptibly further from the place we once were, a gradual ascent that serves to lift us to a place beyond ourselves. Yet our cycle of progress is sawtooth, an asymmetric oscillator that rises slowly only to occasionally drop, in an instant, resetting our standpoint and requiring us to begin again. But this temporary reversal is no setback, simply a mechanism evolved and perfected by nature that, from the countless systems we have already explored, we know helps unite our path with that of others. We entrain ourselves with one another, finding shared purpose and commonality, self-organizing through synchronization so that together we might optimize the systems of our existence in ways that allow ever-greater flow of the prana that sustains us. We take a step back alone, but we then move forward together.

In seeing this wider reality, we are no longer simply objects acted upon by forces beyond our control. Instead we integrate, drawing our attention away from conscious thought and instead toward the heart, finding an inner strength and unity, a freedom in which we become a subject with agency, acting from within. In so doing we engage our unconscious mind more deliberately, develop a higher state of understanding, and in acknowledging the limits of our cognition, learn to aspire to something beyond that which we already perceive. Through states of ecstasy, awe, and wonder, we glimpse a wider field of what might exist, and through that revelation of what is possible, we sing a richer, deeper, and more complete reality into existence. E. F. Schumacher argued for four states of being—mineral, plant, animal, and human—to which could be ascribed the corresponding labels: matter, life, consciousness, and self-awareness. But he also posed the question as to whether there might yet be levels above the human, and that is something that perhaps we might now be able to add. For in our exploration of feedbacks we have seen the cyclic nature of things and the fact that we and all things exist in a learning loop fed from an inexhaustible drive. Above humans, then, is energy, and that energy is information. By stretching our potential beyond the bounds of the visible, into the realm of the ineffable, we connect with the infinite field that shapes the cosmic web. Knowing that energy is what drives us and that energy needs to flow to do work, we reach the inescapable conclusion that "when things are most contradictory, absurd, difficult and frustrating, then—just then—life really makes sense."[31] Why? Because it is in just such moments that we realize that some part of us, some aspect of our life, stands in the way of a flow that cannot be stopped, a mechanism that is "provoking and almost forcing us to develop."[32] Through inward sight we find self-awareness, and in seeing ourselves in the web of a thing much wider and inseparable from our being, we find purpose and meaning.

If we have failed in modern Western society, it is in having reduced learning to simply the acquisition of knowledge in the absence of meaning. Knowledge of the visible world is knowledge of a constrained illusion, a time-lagged prediction composited by our brains based on scant information and a larger amount of preconception. It is a sliver of the wider reality that we inhabit; it is the narrow band of the electromagnetic spectrum that our visual cortex permits us to view. A wider reality—the one of Nobel Prize–winning physicists and Eastern mystics alike—is demonstrably just as real, and it is only by engaging in this extended field of knowledge that we might, if we ask the right questions of ourselves, find what we really need, which is understanding. So although the reductionist sciences are what put our species into space, it was the bigger questions of a wider reality that made us want to do that in the first place: our innate quest not just for knowledge of the lunar surface, but of what it means to be human in an inconceivably vast and empty universe. Our quest is one of never-ending learning, of following the path we have found ourselves upon and that we feel compelled to continue to follow. This search is one we share with every other living being, with each and every quantum of energy in the "void and cloudless sky"[33] that beckons our learning universe to engage in one more loop of advancement, one more step forward, before it steps back. We are connected, we are here, we are a cognitive system absorbing, using, and dissipating a drive. We have reached a state where we now can look inward and choose how we might shape our future and ask what the feedbacks might be that could take us most efficiently to where we need or want to go. Whatever the path we follow, we can be sure that there is something bigger, something beyond just what we can conceive, a flow of information that seeks continuance, survival, persistence, and betterment. And this is a flow that will endure beyond the end of space and beyond the end of time. Because with feedbacks, the end is never really the end.

It's just another beginning, waiting to start.

ACKNOWLEDGMENTS

This book started because the ideas I was exploring for a research article quickly became far too wide-ranging to fit neatly into such a short format. What I didn't know then, of course, was that those ideas were only just scratching the surface of a vast field of truly incredible research that I have since discovered and explored more fully. Spanning diverse disciplines, my labyrinth of investigation led me to places I would have never foreseen and to research that in many cases stunned me with its importance, its clarity, or sometimes both. In drawing together these disparate strands of data, I have been guided only by the desire to reveal the intrinsic beauty of the world we live in and to uncover and describe the simple patterns and behaviors that keep appearing wherever we choose to look. Through reading and researching and attempting to join the dots between widely different fields, my hope has been to share the same sense of wonder that I have always felt, as a scientist and as a human being, when learning about the natural world.

Many, many people have been instrumental in this adventure. My late father, John, nurtured my early interest in the natural world and took me on geological excursions that most likely shaped the direction of my later career. Brian Hopton, one of my high-school teachers, made geology engaging and encouraged me in the right direction when I was clouded with adolescent uncertainty. Of all my university lecturers, Doug Benn was the most captivating and enthusiast scientist I'd ever encountered and single-handedly ignited within me a passion for glaciology that I have retained ever since. David Sugden, my main PhD adviser, taught me the value of scientific inquiry and the importance of accepting when an idea was wrong or incomplete. During the twelve years I spent at the British Geological

Survey in Edinburgh, I worked with many wonderful colleagues, and I am especially grateful to Tom Bradwell, Jeremy Everest, and Andrew Finlayson not just for sharing the scientific journey with me, but for their friendship and steadfast support.

Upon moving to New Zealand I found myself quickly adopted by a new "family," my colleagues and friends at the Antarctic Research Centre. Without them and their incredible passion for research, I would never have found so much enjoyment in the scientific quest. Intimidatingly accomplished as they are, they have always found time to help and guide me in everything I have done there during the last fourteen years.

In tackling the range of topics encompassed by this book, I have stretched my research into areas far outside my training. I am therefore incredibly grateful to the scientists I have contacted for research papers or for clarification of details so that I could better understand and to those who freely gave their precious time to read, comment on, and improve my early drafts: Chris Turney, Tatyana Kulida, Heather Pickard, Tim Stern, James Crampton, Richard Levy, Mario Krapp, Jennifer Hoyal Cuthill, Lisa Miller, David Hayman, Doug Benn, Rebecca Priestley, Priscilla Wehi, Alex Webb, and Dan Cohen. Markus Luczak-Roesch has been an inspiration in the field of complexity science, and I am grateful both to him and to the School of Information Management at Victoria University of Wellington for hosting me for a short sabbatical soon after I began this work. Where appropriate, I have cited about one-third of the thousand or more research articles and long-format sources consulted for this book, but for the sake of readability not every source has been explicitly acknowledged. Similarly, although only a few direct mentions have been provided in the text, I am nonetheless grateful to the authors and administrators of the innumerable online sources I have drawn on when following leads, most especially Wikipedia, NASA, and the Encyclopedia Britannica. The responsibility for any remaining inaccuracies is, however, entirely mine.

One thing I have been especially lucky with during the last decade or so has been the friendship and company of writers such as Alice Miller, Rebecca Priestley, Caoilinn Hughes, and Paul Behrens, who each in their own way have indirectly inspired me to write something of my own. I am especially grateful to the artist Tatyana Kulida for first suggesting that I should write this book and for sharing her philosophical thoughts on all manner of topics over the years. I will always be particularly indebted to my dear friend Heather Pickard, a fellow writer and Antarctican, who witnessed this project grow from its earliest fragmentary ideas, through the years of research and writing, into the finished work it eventually became, all the while providing patient and insightful feedback, much-needed life advice, and relentless encouragement to stay true to my original vision. Jonathan Kurtz was the editor who saw and believed in that vision, and I am truly grateful to him, Nicole Carty, Chloe Batch, Erin McGarvey, and the rest of the team, both at Prometheus Books and Rowman & Littlefield, who helped turn that idea into reality.

Finally, of course, none of this would have been possible without the love and support of my family. I am indebted to my parents, John and Christine, and my brother, Anthony, for everything they have done. My wife, Juliane, has been there every step of this writing journey, quietly but steadfastly supporting me when things were hard and allowing me the space and time to do whatever I needed to do. And so too my children, Frank and Evelyn, who have taught me so much and been a constant source of joy over the years, always asking what I was writing about and whether I was finished yet—thank you, you are wonderful.

GLOSSARY

action potential: When a difference in electrical charge exists between two or more components (or cells), there is a potential for electricity to flow. Whether this flow manifests depends on the nature of the material that separates the charged components. In the human body, the electrical charging of nerve and certain heart cells is configured in such a way as to allow for a relatively slow buildup of energy until a critical state is reached. At this point, rapid release of the accumulated charge drives some form of action—a nervous impulse or contraction of the heart muscles, for example.

albedo feedback: The amount of solar radiation either absorbed or reflected by the Earth's surface can be measured, and that number is referred to as its albedo. This number varies between zero and one and represents the proportion of incoming energy that is reflected. The Earth has an average albedo of 0.3 (or 30 percent), but absorbent areas such as forests, oceans, and dark rock have values of around 0.1 or lower, whereas ice and snow have a much higher albedo of 0.5 to 0.9. As the amount of snow and ice on the Earth's surface changes through time, it exerts an extremely powerful feedback on the climate, amplifying initially small temperature changes into major climatic fluctuations.

allometry: The concept that an organism's characteristics, or the processes controlling its form or behavior, relate to one another by way of a scaling law. For example, species within a genus may vary in size, but in some cases may share similar proportions, or scaling relationships, in their body plan.

alpha wave: Brain cells, or neurons, use electricity to relay signals between each other. If a group of neurons emits a coordinated, or synchronized, electrical signal, the oscillation in activity is detectable using an electroencephalogram. The frequency of the electrical wave determines its type—alpha waves represent oscillations of around 8 to 12 hertz, or cycles per second. Alpha wave activity is usually associated with being in a relaxed but wakeful state. *See also* theta wave.

Antikythera mechanism: Discovered in 1901, this two-thousand-year-old device was the first mechanical computer, using bronze gearwheels to track planetary movements. Attempts to understand how precisely this was achieved have been hampered by the degraded state of the eighty-two remaining fragments, but the latest (electronic) computer reconstructions of this machine reveal fascinating insights into just how capable its Greek inventors were.* The challenge of tracking the ever-changing positions of the sun, moon, and the planets Mercury, Venus, Mars, Jupiter, and Saturn relates in part to their very different orbital periods (from Mercury's 88-day rotation around the sun, to the 10,747 days of Saturn's). Given large enough wheels, gearing ratios could of course be found to almost exactly match the required revolutions of each orbital cycle, but such a device would be excessively large. Instead, the genius of the creators was to find the smallest common factors that allowed the periodicity of each orbiting body to be captured as accurately and in as compact a form as possible. By combining thirty-seven differently sized cogs, the largest of which hosting 223 teeth and measuring only 13 centimeters across, sufficient gear combinations could be realized to enable an economy of construction that nonetheless still allowed planetary wanderings, phases of the moon, and the timing of eclipses to all be predicted precisely. The Antikythera mechanism is not just a marvel of construction and ingenuity; it is the first known device that facilitated the mechanization of astronomical predictions.

Apollo; Apollo missions: The Greek god Apollo was the son of Zeus and Leto and the twin brother of Artemis. He was the god of music and poetry, light, healing and prophesy, and established the Oracle of Delphi where mortals sought divine wisdom. During the era of space exploration, the idea of Apollo as the charioteer god moving the sun through the sky was adopted by the US space program and used as the name for a series of missions from 1961 to 1975. It was the *Apollo 11* mission that achieved the first crewed lunar landing in July 1969.

---

* Freeth, T., Higgon, D., Dacanalis, A., MacDonald, L., Georgakopoulou, M. & Wojcik, A. A model of the cosmos in the ancient Greek Antikythera mechanism. *Scientific Reports* 11, 1–15 (2021).

archaea: One of the three domains of life on Earth, archaea are single-celled organisms that lack a cell nucleus; that is, they are prokaryotes. Archaea are distinct from bacteria and eukarya, the other domains of life.

Archean: The geological eon spanning the period four to two-and-a-half billion years ago. There were no continents at this time, and Earth was mostly covered by vast oceans.

artificial intelligence. *See* machine learning.

assembly theory: Originally devised by Leroy Cronin and later developed by others, such as physicist Sara Walker, this theory proposes that molecular complexity can be defined by a single number—the assembly index—which describes the number of steps that are required to build a certain structure from sequentially assembled parts. In structures where patterns repeat, sequential assembly allows the repeating pattern to be duplicated without further innovation, essentially allowing for a shortcut. But when structures have no repeating elements, shortcuts are impossible. The latter have a much higher assembly index and a much lower probability of being created randomly. Systems with high complexity and high assembly indices are therefore assumed to imply creation by a living organism, rather than by chance combination.

associative learning: The idea of "unlimited associative learning" has been proposed as one mechanism by which complex life evolved and continues to evolve. In this theory, an organism's ability to find associations between disparate pieces of information and to learn from these associations in ways that guide its future behavior dictates that organism's ability to survive and proliferate.

Australopithecus: A genus of early hominin from the Late Pliocene, from which other genera—such as our own *Homo*—subsequently evolved.

awe: When we are unable to fully accomodate an experience within our existing framework of knowledge and understanding, we become cognitively overwhelmed. In instances when the experience is one of large physical scale or intense emotion, we might call this a state of awe.

basal ganglia: A brain region that sits below the outer layer, the cortex, and is not only associated with control of voluntary body movements but also with learning, cognition, and emotion. The basal ganglia is heavily involved in both tool use and language skills.

biodiversity: A measure of the genetic diversity of different organisms in an environment, often seen as a measure of ecological health. A more biodiverse environment is typically a more resilient and productive one. Loss of global biodiversity due to human activities such as industrialized farming, deforestation, and urbanization is currently a major concern for the viability of our own species.

biomass: The combined mass of material that makes up organic life, sometimes expressed in terms of its weight of carbon (since all living things on Earth are carbon based). Biomass is a useful measure of the amount and distribution of life in different environments.

Cambrian explosion: Around 541 million years ago, a rapid proliferation of diverse forms of life appears in the geological record, marking the permanent transition of life from the oceans onto land. Over only a few tens of millions of years, the surface of the Earth became far more densely and diversely populated, enabling the rate of biological evolution to accelerate enormously.

carbon dioxide greenhouse effect: Although Earth's atmosphere contains on average only 420 parts per million (ppm) carbon dioxide, this gas exerts an extremely strong effect on Earth's temperature. It does this by absorbing long-wave radiation that is emitted from land and ocean surfaces and then re-radiating that heat in all directions, including back toward Earth. Small changes in the concentration of atmospheric carbon dioxide therefore lead to significant changes in surface temperature. For example, when the climate warmed by around 6 degrees Celsius over the ten thousand years from the end of the last ice age to the mid-nineteenth century, carbon dioxide concentrations rose only 100 ppm. Since that time, concentrations have further increased another 140 ppm, of which only 1.1 degrees Celsius warming has yet been realized. The full consequence of human-caused carbon dioxide emissions since the early industrial era will only be felt later this century and beyond.

Carboniferous: A geological period of the Paleozoic era spanning roughly 359 to 299 million years ago. During this time, Earth's surface was initially warmer than present due to relatively high levels of atmospheric carbon dioxide, but as concentrations declined, the climate cooled. The warm and humid climate in tropical regions fueled lush plant growth across many areas of exposed land, as well as extensive swamps. These swamps preserved fallen vegetation and gave rise to extensive coal beds.

cerebral cortex: The outer layer of the brain. In humans, the cerebral cortex contains approximately fifteen billion neurons.

chaos, nonlinearity: The way that many dynamical systems evolve through time depends on their state at the beginning, the so-called initial conditions. In linear systems, changing the initial conditions by a certain amount tends to lead to changes in the result by a corresponding and predictable amount. But in chaotic or highly nonlinear systems, a small change to initial conditions can lead to vastly different outcomes, due to the strength and sensitivity of feedbacks within the system. Weather is a good example.

chemosynthesis: Whereas photosynthesis is the process by which organisms such as plants convert light energy into fuel to facilitate growth, chemosynthesis is the process by which growth is fueled directly from the energy liberated from chemical reactions. Organisms that use chemosynthesis are commonly found in locations where light is absent and reactive chemicals are abundant, such as in the water surrounding deep sea volcanic (or hydrothermal) vents.

cladistics: A method of categorizing species based on shared traits. By identifying relationships between organisms in this way, an evolutionary tree can be constructed and common ancestors identified.

cognition; cognitive cycle: In its simplest form, cognition can be thought of as learning. Any system that changes its behavior in response to the result of previous actions can be considered a cognitive system. Cognition does not imply consciousness, or intent, only an ability to gradually optimize.

collective intelligence: The combined cognitive abilities of a group of people may, under certain circumstances, produce a distributed form of intelligence that guides group behavior and learning. By bringing together the wisdom of a large number of individuals and by providing a means by which individual ideas can be shared and assessed, collective intelligence has the potential to drive rapid change in a system and the emergence of novel behaviors.

complex system; complexity theory: A complex system is composed of multiple components, each of which may be governed by its own forces but which is also coupled in some way to other components of the system. Changes to any one of the components will affect other parts of the system, leading to systemwide changes as a cascade of feedback is transmitted from one component to another. Complexity theory uses insights from observation, experimentation, and modelling to find general rules that govern the evolution of complex systems.

consciousness: Research during the last two decades has provided us with an increasingly useful framework for defining consciousness. According to the information integration theory, consciousness arises from an organism's ability to bring together disparate threads of information in ways that allow it to make predictions that can guide its behavior and improve its survival outcome. An implication of this theory is that consciousness is not a single quality, but rather a phenomenon that exists as a continuum of possible states, from minimally conscious to fully self-aware.

controlled hallucination: We do not experience reality as it actually is, instead we construct our own representation of reality based on incoming stimuli that we contextualize within a framework of our own expectations and memories. We form a picture of the world, a hallucination, and then progressively refine (control) it to account for the continual stream of information we absorb.

craton: A long-lived portion of a continental lithospheric plate, typically preserving older rock than is found surrounding it.

criticality: Stochastic (random) events can push a complex, nonlinear dynamical system that was sustainably oscillating around a point far from equilibrium into a chaotic state that results in abrupt change or even complete collapse. Systems that are close to this point are deemed critical, and the increasing amplitude of fluctuations that precedes the abrupt change is termed *critical slowing down*, because each oscillation takes longer and longer to regain a stable state. Critical systems often exhibit characteristic statistical properties, such as power-law relationships. Such statistical distributions are often called *1/f noise*.

critical transition. *See* tipping point.

cyclicity: Processes or quantities that vary through time in predictably oscillating ways are considered to be cyclical or periodic. The magnitude of each oscillation may differ, but where the spacing between each peak is relatively constant, such signals are said to exhibit a characteristic frequency, or periodicity. Identifying robust cyclicities in natural phenomena can be difficult, but if found, they can provide useful clues as to the processes that underpin those phenomena.

default mode network (DMN): A collection of brain regions that form a functional, rather than structural, network. The DMN is active when our brains are not occupied with specific tasks and is often referred to as the "task-negative network." The DMN is associated with inner thoughts, brooding about the past, and daydreaming or worrying about the future. An impaired ability to switch off the DMN is typical in people suffering from depression and anxiety, whereas sufferers of attention deficit hyperactivity disorder (ADHD) tend to have higher activity in the DMN at the same time as the "task-positive" or attentional brain networks are also active. This makes it difficult to focus on any one task at a time, and people with ADHD may have a tendency to become easily distracted.

dissipative adaptation: Developed extensively by physicist Jeremy England, this is the idea that a collective of interacting elements in a given system will absorb energy from an external "drive" and direct (or *dissipate*) that energy in ways that are determined by the structure of the system. If one part of the structure absorbs, rather than redirects, energy, the overall flow of energy is impeded. In many cases, this loss of efficiency triggers adaptations to the structure that improve the flow of energy through it, so that over time, the system as a whole becomes preferentially aligned with the energetic environment in which it exists.

dissonance: When tension exists between components in a system, there is discord. This may be, for example, in the form of interpersonal relationships (social dissonance) or in the way that we struggle to come to terms with events that do not conform to our expectations (cognitive dissonance). Dissonance tends to lead to a feeling of unease, prompting us to find a way to resolve the conflict at the source of the dissonant experience.

dynamical systems theory: A dynamical system is one whose evolution through time depends on its previous state. This contrasts with a purely deterministic system that changes through time according to conditions that remain unchanged since its initiation. A single swinging pendulum could be thought of as a deterministic system, whereas a system composed of many pendula, each influencing each other, would be a dynamical system, because the feedbacks between each pendulum make the evolution of the system harder to predict.

ecological niche: All organisms, regardless of their level of cellular or metabolic complexity, have a preferred environment to which they are best adapted. This is their ecological niche, and it is usually distinct in some way from those occupied by other species. Overlap between niche spaces leads to competition, most likely resulting in adaptation by one or other species so that the overlap is reduced.

electromagnetic spectrum: Electromagnetic radiation pervades the entire cosmos. Different components of this radiation can be measured in terms of their wavelength. Visible light, for example, has wavelengths in the range of approximately four hundred to seven hundred nanometers, but other forms of radiation, such as gamma waves, have wavelengths a trillion times shorter, whereas radio waves can have wavelengths trillions of times longer.

emergence: In complex systems involving interactions between multiple components, novel behaviors may arise that are unanticipated or hard to predict. The emergence of life from a small number of chemical elements, for example, is a phenomenon that remains difficult to explain.

eukaryotes: In contrast with prokaryotes, these are organisms whose cells contain a nucleus. By incorporating compartments within a cell, eukaryotes allow different chemical reactions to be carried out in isolation from one another, enabling a single cell to undertake a greater range of discrete tasks. As such, eukaryote organisms are far more adaptive and complex than prokaryotes, and today they make up 99.5 percent of life on Earth.

evolutionary decay; evolutionary decay clock: A concept coined by Jennifer Hoyal Cuthill and others in their 2021 paper in which they examined the fossil record and showed that at certain times in the geological past, species co-occurrence dropped to very low levels and that these changes were cyclical. During such times, the rate of evolution slowed or decayed.

feedback: The mechanism by which the consequences of an action or process influence another action or process. Feedbacks may be positive, driving further change in the same direction, or negative, tending to suppress or reverse change. Coupling between components of a complex system may lead to a cascade of feedbacks that radically and rapidly alters the state or behavior of the system as a whole. *See also* phase transition.

Fermi paradox. *See* Great Filter.

fractionation: The process by which a chemical compound fully or partially separates into constituent parts.

functional networks: Regions of the brain, even if they are not physically adjacent, may be electrically connected in ways that form networks. Each network of brain regions controls a specific set of information-processing tasks. Depending on the complexity of these tasks and their importance for basic functioning, the functional relationships form either "primary" or "higher-order" networks.

functional redundancy: The ability of different species to serve the same functional roles in an ecosystem, such as breaking down leaf litter on a forest floor or aerating the soil by burrowing. These functions may be essential for the survival of other organisms. The more species that serve similar roles ("redundancy"), the more resilient as a whole an ecosystem will be. *See also* mutualism.

fungal filaments: Fine threads of fungi in soil (hyphae) provide invaluable assistance to plants by binding with their roots and helping deliver nutrients and water to the plant. In some instances, these hyphae also play a vital role in transmitting electrochemical signals between individual plants, enabling an entire population to communicate with one another.

Gaia; Gaia hypothesis: In Greek mythology, Gaia was one of the primordial gods and represented both the Earth and the life upon her. In more recent times, Gaia has become associated with the concept that the Earth and all its constituent parts behave as a single, living organism. The Gaia hypothesis put forward by James Lovelock and Lynn Margulis in 1974 argues that once life had evolved on Earth, it took hold of the atmosphere and, through a series of feedbacks, modified it in ways that ensured life's persistence.

gigantism: The abnormally large growth of individual organisms or species is common throughout the fossil record of the last 541 million years and is most commonly associated with changes in environmental conditions that favored extreme growth. This may have been in the form of higher levels of atmospheric oxygen or changes in temperature. In some species it is also apparent that body size naturally increases through time; that is, individuals of successive generations tend to be larger than their predecessors.

glacial-interglacial cycle: Throughout Earth's history, the climate has fluctuated between cold and warm states, giving rise to periods when glaciers and ice sheets advanced or receded. During the Cenozoic era of the last sixty-six million years, oscillations between glacial (cold) and interglacial (warm) climates have become increasingly regular. The orbit around the sun leads to slowly varying changes in solar energy that exert a key control on Earth's climate, and over the last one million years these changes have progressively shaped it into one characterized by periodic one-hundred-thousand-year cycles of alternating warm and cold states.

Great Filter: The physicist and Nobel laureate Enrico Fermi posed the question as to why, in a universe as vast as ours where extraterrestrial life should be abundant, have we so far been unable to detect signs of it? This paradox can be partially satisfied if it is considered that there may be myriad ways in which life, even intelligent life, can be extinguished. Given the relentless certainty of any given planetary home being bombarded by extraterrestrial impacts, solar radiation, or experiencing catastrophic climate change, mass extinction, or nuclear warfare, the chances of life on those planets surviving successive "filters" long enough to be detected by life elsewhere in the universe might be vanishingly small.

Great Oxidation Event (GOE): As early life took hold in the ocean, mostly in the form of phytoplankton, it produced increasing quantities of oxygen. This oxygen dissolved into the seawater, but over time, pulses of oxygen escaped to the atmosphere, where it was readily taken up by reactive chemicals at Earth's surface. The GOE records the consequence of these successive pulses, spanning approximately two-and-a-half to two billion years ago.

Hadean: The geological eon that includes the first 530 million years of Earth's history, from 4.53 to 4 billion years, and also includes the oldest solid material in our solar system, dated to approximately 4.57 billion years. The Hadean Earth was characterized by a dense atmosphere rich in carbon dioxide.

Homo; Homo erectus; Homo sapiens: From Australopithecus emerged a new genus, *Homo*, around two million years ago. This group of great apes began with *Homo habilis* and subsequently gave rise to other species such as *Homo erectus*, *Homo neanderthalensis*, and modern humans, *Homo sapiens*.

hydrothermal vent: Fissures on the seafloor, where heat from buried magma brings hot water to the surface, often form conical vents as dissolved chemicals precipitate out of solution. Over time, "chimneys" develop that support a unique and exotic ecosysten tolerant of hot and chemically rich environments.

hyphae. *See* fungal filaments.

ice core: Snow that falls in areas cold enough to avoid seasonal melting accummulates layer by layer, gradually becoming compressed under its own weight. In places like Antarctica and Greenland, ice sheets have built up in this way over hundreds of thousands of years. Because the water comprising the ice crystals that initially fell as snow contains slight isotopic variations, depending on the temperature and atmospheric conditions at the time it was formed, each layer of ice can be analyzed to determine details of past climates. To recover continuous cores of this ice, drill rigs are taken to the coldest and highest parts of an ice sheet and bore down vertically into the ice using a hollow barrel, so that the sampled ice rises up through the center.

ice sheet. *See* sea ice.

ice shelf. *See* sea ice.

information integration theory (of consciousness): This idea, proposed by Guilio Tononi, argues that consciousness depends on the ability of a system to integrate information. A greater capacity for information processing through the association of disparate bodies of information allows for a higher level of consciousness. Our state of consciousness, therefore, depends on how well we can assimilate and find connections between incoming streams of information.

intermittent fasting: Depending on the energy density of their preferred foods, animals may eat almost continually during waking hours (such as grazing herbivores) or episodically (such as predatory carnivores). As omnivores, humans are able to chose what and when to eat. Experiments have shown that incorporating periods of fasting into our daily routine leads to many cognitive and physiological benefits, slowing aging and helping us to maintain a healthy weight.

International Geophysical Year: In an attempt to reignite global scientific collaboration after the so-called Cold War, the eighteen-month period from July 1957 to December 1958 was designated as a time for renewed collaborative research. The focus spanned a wide range of Earth science disciplines, such as meteorology, glaciology, and geology. Sixty-seven countries took part.

interoception: In addition to the continual stream of external stimuli received by the brain—sensory inputs arising from environmental conditions outside of our physical body—we also generate stimuli from within our body, from which our brain gauges our current state of health or well-being. Subtler than external stimuli, internal messages are transmitted from body to brain and vice versa through a vast network of nerve fibers, many of which are integrated into the vagus nerve. Interoception is the ability to be aware of these internal messages.

IPCC: The Intergovernmental Panel on Climate Change, a United Nations–mandated advisory body tasked with providing periodic assessments of the most recent climate change research. Composed of three working groups (physical science, impacts, adaptation), the IPCC produces extremely detailed and thoroughly peer-reviewed reports every seven or eight years. These reports are used by governments all over the world to help guide policy.

isotopes: Differing forms of the same chemical element, in which the number of protons in the atomic nuclei are identical but the number of neutrons differs.

learning loop: The process of learning often relies on a cycle of experimentation and assessment, deliberate adjustment, and then new experimentation. Such learning loops are able to drive purposeful change in almost any circumstance. In humans, we might identify small learning loops that are individual or personal in nature and large learning loops that require societal or national-level modification and change.

limit cycle: In a closed system, feedbacks between dominant components can, under certain circumstances, result in a fluctuating system that periodically oscillates between two equally stable states. *See also* predator-prey system.

Linnaean classification: In 1735, Carl Linnaeus published a scheme that allowed life on Earth to be classified in a rigorous and systematic way. Each organism was identified by two names—its genus followed by its species name. This became known as "binomial" classification and has persisted in use for nearly three hundred years.

lithospheric plates: Fragments of the rigid lithosphere that form coherent blocks and move as a single unit. Plates interact with one another at plate boundaries, where they might slide past one another sideways, pull apart and allow new rock to rise to the surface, or push together. Where two plates of similar density push against one another, they crumple to form mountain ranges such as the Himalayas. Where plates of differing densities collide, the denser one is pushed down (subducted) beneath the less dense one, forming an oceanic trench such as the Mariana Trench in the western Pacific Ocean.

machine learning: The Newtonian scientific method relies on taking a small amount of observational data and devising, then testing, ideas that can explain those data. Increasingly, however, our digital lifestyles result in vast amounts of data that are near impossible to analyze with traditional methods. Machine learning is a type of artificial intelligence that looks for patterns in large amounts of data, usually by employing statistical techniques such as regression. By analyzing huge volumes of data with regression statistics, a computer program can deduce the underlying relationships between a set of variables and then make predictions for specific combinations of those variables that might be of interest. Machine learning and artificial intelligence now underpin many facets of our lives, from simple tasks such as suggesting restaurants based on our likes and dislikes, to more complex challenges that involve real-time human-computer interaction via online chatbots that try to answer our questions.

magnetic field: The Earth's iron core and convecting mantle give rise to a geodynamo effect, which produces a magnetic field. This field has poles that wander over time as a result of internal Earth processes, but typically lie close to the geographic north and south poles. Periodically, the strength of this field can weaken to the point where the poles are indistinct. Repolarization may then take place in a way that reverses the previous orientation of the field. Magnetic reversals have taken place repeatedly throughout Earth's history, but it remains a mystery as to why these events happen when they do.

mantle plume: An elongate, approximately vertical, structure in the Earth defined by hot mantle rock that rises more rapidly than surrounding material due to its greater buoyancy. Mantle plumes may bring molten rock to Earth's surface and produce volcanic islands. Plumes may also play a role in the movement of lithospheric plates.

memory: Retention and recall of information is a critical part of consciousness and of learning. Whether memory takes the form of information encoded in a mammalian brain or in the form of a particular structural configuration that enables an abiotic process to reproduce a previous pattern of behavior, all cognitive systems, regardless of their complexity, benefit from the accurate use of memory.

Mesolithic: The "middle stone age" marks the period from the end of the last ice age, around 11,500 years ago through around 5,000 years ago, when it was supplanted by the Neolithic, or new stone age. The Mesolithic period was characterized by more advanced stone-working techniques than was evident in Paleolithic societies but lacked the sophisticated technologies of the Neolithic.

methane hydrate: Methane is a carbon-based compound that is a gas under the atmospheric pressure and air temperature across most of the Earth's surface. It is often formed as organic (plant) material decomposes. However, if plant material is buried in saturated sediment far enough below sea level, the methane that is produced can take a crystalline hydrate form. The methane is trapped in this form as long as low temperatures and the pressure from above is maintained.

microbial loop: A three-part cycle in which protozoa, phytoplankton, and bacteria are connected through a continuous cycle in which they each benefit from nutrients released from, or absorbed by, one of the others.

microbiome: A distinct population of microorganisms, such as fungi, bacteria, and viruses, that occupy a particular habitat. Animals host populations of microbes in areas such as on their skin and in their gut, which, through contact or inhalation, interact directly with the microbiota of the soil or other organisms.

Milanković theory: Around one hundred years ago, the Serbian scientist Milutin Milanković quantified the amount by which the solar radiation reaching the surface of the Earth varied through time as a consequence of three different orbital variations. Over time periods of around twenty-three thousand years, the wobble of the Earth on its axis leads to a process known as precession, which affects the way that incoming radiation is distributed over the globe. The amount of tilt of the Earth's axis, known as its obliquity, also affects the pattern of solar energy, but over approximately forty-one thousand years. Over even longer time-scales, the amount of energy reaching the Earth varies depending on how circular or elliptical our orbit around the sun is, leading to changes in radiation over one hundred thousand and four hundred thousand years. Building on previous work, Milanković's orbital theory convincingly explained the causes of ice age cycles on Earth.

mindfulness: The practice of being present, focusing only on what one is experiencing or feeling at a given moment and letting go of distracting thoughts that might dwell on the past or anticipate future events. Meditation that employs mindfulness often allows a practitioner to find inner calm.

Miocene, Pliocene, Pleistocene: Three geological periods of the Cenozoic era, collectively spanning the last twenty-three million years during which the global climate slowly cooled, global vegetation patterns adjusted to drier conditions, and the "modern" configuration of tectonic plates, as well as regular ice age climate cycles, became established. It was during this time that humans evolved from our primate ancestors.

mutualism: Species that depend on one another for services may be considered symbiotic. Mutualism is one type of symbiotic relationship in which members of the same, or different, species, all benefit from their close association. *See also* functional redundancy.

Neolithic: Literally, the "new stone age," an archaeological period characterized by advanced stone tool use before the advent of metalworking. Neolithic cultures and the technologies they employed appeared at different times in different places, commonly arising around five thousand years ago and lasting for several thousand years.

Neoproterozoic: The geological era that spanned the last half-billion years before complex life emerged en masse during the Cambrian explosion, 541 million years ago.

Neoproterozoic oxygenation event: The second of two main phases of global oxygenation that took place during Earth's geological past. Likely occurring as a series of pulses, rather than a singular event, the Neoproterozoic oxygenation event allowed for much higher concentrations of atmospheric oxygen than had resulted from the earlier Great Oxidation Event, a transition that most likely played a critical role in the emergence and proliferation of life on land 541 million years ago.

network; network theory: Any system can be considered as a network of interconnected nodes, or elements. Computational network simulations can then be run in which connections are either added or removed according to some predefined rules. Through this kind of on/off switching, networks can be made to evolve dynamically and independently, as changes in one property of the network feed back on other parts of the system. And because the statistical properties of a network can be easily calculated, experiments can be run that allow very precise measurements of the circumstances surrounding any changes that take place.

orbital variability. *See* Milanković theory.

orogenic thickening: At collisional plate boundaries, the pressure of persistent tectonic force leads to compression and shortening of the lithosphere. To accomodate this shortening, the compressed rock crumples, pushed both downward and upward. Above ground, this uplifted rock forms mountain belts. Below ground, the thicker lithosphere forms a "keel" that supports the mountains above.

oscillator: In physics, any component whose behavior fluctuates between two extremes, or end-member states, can be considered an oscillator. Such fluctuations may be in its energy state, perhaps in terms of its electrical charge or its potential energy. Oscillators may be coupled, in which case energy is transferred from one component of a system to another.

Paleolithic: Spanning the late Pliocene approximately 3 million years ago through the end of the Pleistocene 11,500 years ago, the Paleolithic ("early stone age"), encompasses almost the entire period of hominin stone tool development and technological discovery. Later periods, the Mesolithic ("middle stone age") and Neolithic ("new stone age"), were much shorter in duration but saw far more rapid development in terms of stone-working skills, building construction, and social organization.

parietal lobe, fronto-parietal network: The parietal lobe is one of the four main brain regions, located between the frontal lobe and the occipital lobe and above the temporal lobe. This brain region and the network that it forms with its neighbors is responsible for the way in which external (i.e., sensory) stimuli are processed and interpreted. It is essential for navigation and for language processing and may be at least partially responsible for our "sense of self."

periodicity. *See* cyclicity.

phase transition: When a system is undergoing change, it might reach a point at which feedbacks within the system lead to rapidly accelerating change that mutually reinforce one another. From an initially slow start, this accelerating change leads to an exponential response until the system is saturated, that is, the point at which no further change is possible, or the rate of change dramatically slows. The period of rapid change may lead to such widespread changes in a system that it now exists in a different state entirely. This is a phase transition, or in ecological contexts, a regime shift.

planetary intelligence: Some theorists propose that processes taking place on Earth, such as the continual reshaping of its tectonic plates, evolution of its climate, and genetic turnover of species, are all part of a cognitive, or learning, process. Where novel configurations or behaviors prove successful, they will proliferate more widely than those that are less useful and thus will be retained. Over time, the entire planetary system will evolve toward a state that is most optimally suited to the external conditions that govern it. This process of refinement can be thought of as an expression of intelligence.

Pleistocene. *See* Miocene, Pliocene, Pleistocene.

Pliocene. *See* Miocene, Pliocene, Pleistocene.

polarity, magnetic. *See* magnetic field.

prana: A Sanskrit word from yoga and Indian medicine that translates variously as the "vital" or "life force" energy not just of living beings, but of the entire universe. It can be thought of also as the "breath of life." Accordng to Hindu texts, there are five types, or "winds," of prana.

predator–prey system: In a theoretical isolated environment, species may interact with each other competitively, such that their fates are intimately entwined with one another. When one species preys on another, survival of the predator depends on not exhausting the available prey. Overexploitation leads to a collapse of the prey species and a lack of food for the predator, leading to a reduction in its numbers too. This allows the prey to recover and so on. Often such systems might evolve to reach a stable state in which the population size of both species oscillates around a sustainable balance point. The concept can be applied more conceptually to any system of coupled or interdependent components.

prediction: The process of assimilating observations with other stimuli and existing knowledge to make an informed guess regarding future events. Because a prediction, in contrast to a guess, is based on the integration and interpretation of disparate sources of information, it may be updated incrementally as new information becomes available.

prokaryotes: Single-celled organisms lacking a nucleus, distinct from eukaryotes. Prokaryotes are limited in their functional capacity and so cannot achieve complex forms. Today, only 0.5 percent of known species are prokaryotes.

proxy record: Natural recorders of changes in climate can be interpreted to reveal what past atmospheric and oceanic conditions were like, for times before direct measurements were possible. These archives might be in the form of coral reefs, ice cores, or sediment cores from the ocean. In each case, the chemical composition of the materials of which they are composed can be analyzed to determine the conditions under which they were first formed.

psychedelics: From the root words *psyche*, meaning "mind," and *dēlos*, meaning "clear" (the latter from the same root as derivatives such as deity, or deus, pertaining to a heavenly spirit or god). The term was first used by Humphrey Osmond in 1957 to describe a class of compounds whose properties led to consciousness-altering or hallucinogenic experiences.

quantum uncertainty: In essence, Heisenberg's uncertainty principle asserts that there is a limit to the accuracy with which the velocity and position of a subatomic particle can be predicted. Accurately measuring velocity, for example, means that the position is hard to define and vice versa. Where quantum mechanics differs from classical mechanics is that with the latter, an object can only exist in one state at a given time, so its properties can be measured easily. With the former, however, a particle may exist in a superposition of different states; that is, it can be in two places at once.

regime shift. *See* phase transition.

reversal, magnetic. *See* magnetic field.

reward circuit: The mesocorticolimbic circuit is a network of brain regions that collectively motivate animals to behave in ways that enhance their chances of survival, such as eating or learning, or the chances of survival of their species, such as having sex. The "reward" for undertaking such activities is the release of the neurotransmitter dopamine, which produces a pleasurable sensation in our brain and thereby encourages more of the same kind of activity.

scale invariance: A statistical relationship in which an order-of-magnitude change in one variable leads to approximately the same scale of change in another variable, regardless of whether the quantity is large or small. Also known as "scale-free behavior," these kind of relationships are often described by power-law relationships, and can be shown as straight line graphs when plotted with two logarithmic axes.

sea ice: Across the vastly different cold environments of the world, ice can take many forms. Where the ocean surface freezes, salt (brine) is rejected during freezing and the less saline component freezes into flat plates of sea ice. By contrast, an ice sheet is composed of ice formed from snow accummulating on land that becomes increasingly compressed by its own weight, eventually becoming so thick that its starts to flow. Where an ice sheet flows into the ocean, it starts to float, forming an ice shelf.

self-organization: Through internal feedback processes that incrementally modify the behavior of individual components of a system arises an increasing degree of synchronization within the population as a whole. When the entire distributed population of components are acting in a coordinated manner, the system as a whole can be considered to have reached a self-organized state.

sentience: The ability of an organism to respond emotionally to sensory stimuli, such as pain. In recent years, the list of sentient creatures has grown considerably, as we become increasingly able to differentiate emotional responses in nonhuman species from purely physical reactions.

snowball Earth: Periods of geological history when a large proportion of the Earth's surface was covered with ice or snow. Typically, glaciers extended to low latitudes and to elevations much lower than is possible under today's climate regime. Distinct from the ice age cycles of the recent geological past, "snowball Earth" events were characterized by widespread ice-cover across the oceans as well as the land, leading to far greater reflectance of solar energy—a positive feedback that maintained the glacial state. Several such events are thought to have occurred, from the Huronian glaciation more than two billion years ago to Neoproterozoic glaciations less than one billion years ago.

social capital: A way of measuring the extent to which an individual benefits or acquires some form of advantage from their network of connections and wider group associations.

social resilience: The ability of a society, or coherent population, to persist as an identifiable group in spite of external or internal pressures or stressors. These could be environmental change, habitat destruction, natural disasters, famine, drought, or disease. In human societies, social resilience may also refer to our ability to withstand economic or political upheavals.

socioeconomic impact: Events such as the Great Depression of the 1930s were social impacts of a largely economically precipitated shock. This arose largely from a single chain of related events. In 1920s

America, rising optimism during the postwar years and the emergence of technological innovations fueled widespread investment and steady economic growth. As investor confidence grew, stock values steadily increased, reaching a peak in the summer of 1929. But by then, early signs of an economic contraction had already started to appear. Faced with the risk of owning devalued stock, investors began to sell, setting in motion a wave of panic that swept through the market, each sale triggering more and more desperate offers. The Dow Jones index—the price-weighted measure of thirty listed companies used as the yardstick of stock market value—fell 23 percent in two days, ultimately bottoming out three years later, having lost 89 percent of its previous value.

space-time continuum: Before the early part of the twentieth century, the geometry of the universe was measured in the three dimensions that we most easily comprehend, and time was considered as something separate. But in his theory of relativity, Albert Einstein proposed that the perception of passing time itself depended on how fast an observer was moving relative to the object being observed. As a result, space and time become conjoined into a single four-dimensional mathematical frame of reference.

specialization trap: The evolutionary pressures that drive genetic and behavioral adaptation of an organism can result in a species becoming so well suited to a particular ecological niche that they are unable to easily adapt to new circumstances—they become "trapped" by their degree of specialization.

speciation, radiation, extinction, turnover: The process of the formation of new species (speciation) and the spread of these new species through the genetic landscape (radiation) leads to an overall increase in genetic diversity in a population. Extinction, on the other hand, removes species from the gene pool and reduces diversity. The rate at which species arise and go extinct can be thought of as the rate of genetic turnover.

species-area relationship: Where life is free to evolve in a given space, it tends to result in not just more individuals in a larger space, but also a greater number of species. The relationship between the number of species in an area of a given size typically follows a power-law distribution; that is, it is scale-invariant.

supercontinent: Following the initial fracturing of Earth's lithosphere, the emergence of plate tectonics, and the creation of continental rocks, continued motion of the plates periodically brought continental masses together in ways that formed much more extensive land areas. Some of the largest and best reconstructed supercontinents were Pangaea (336–175 Ma), Gondwana (550–175 Ma), Rodinia (1,130–750 Ma), and Nuna (1,820–1,350 Ma).

superposition: *See* quantum uncertainty.

symbiosis. *See* mutualism.

synchronization: The mechanism by which independent components of a system begin to act in unison. In social animals such as humans, this may include motoric synchronization, in which we adjust our own movements or behaviors to mirror those around us.

Tao, Taoism: Taoism most likely arose with the writings of the Chinese philosopher Lao Tzu, around the fifth century BCE. As a concept, the Tao is loosely defined but is typically considered to represent the origin of the universe and the fundamental essence that pervades all things. Living in harmony with the Tao is therefore the primary goal of the Taoist.

tectonic plates. *See* lithospheric plates.

theory of mind: When conscious, we might be aware of our own thoughts. And by extrapolating from our own experiences, we might also assume that other beings like us also have thoughts. It naturally follows from this assumption that we might then think about what others are thinking or that they might wonder about our thoughts.

theta wave: Like alpha waves, theta waves are produced by electrical activity in the brain. They may arise from the hippocampus or the cerebral cortex and, at 4 to 8 hertz (cycles per second), are of a lower frequency than alpha waves. Theta activity increases in very relaxed states, such as waking from sleep or during deep meditation. Theta waves enable neural synchronization and information exchange, even between individuals (such as a mother and child) and play an important role in memory formation.

tipping point: The kind of nonlinear behavior taking place during catastrophic events like a stock market crash or the collapse of a natural system can often be retrospectively tied to a point in time or to a series of closely related events when the normally slow and steady change in one direction changed to one that was either starting to accelerate or starting to move in a different direction altogether. That point of inflection is a phenomenon known as a tipping point, or critical transition.

transcendence, transcendental state: Typically, the idea of transcendence is related to a spiritual or religious experience that manifests as something beyond or above the ordinary. Likely associated with a feeling of awe, a transcendent experience may be one that is difficult to comprehend in the context of our daily lives, and to understand it may require an acceptance of unknown forces. Transcendental Meditation practitioners deliberately seek to attain a transcendental state in which alpha wave activity increases and activity in the right and left temporal lobes becomes much more coherent.

triple junction: Where three tectonic plates are adjacent. These are often, but not exclusively, found in areas where the lithosphere is fracturing and breaking apart; that is, the plates are diverging.

volcanic arc: An elongate distribution of volcanic vents created by lithospheric subduction and melting.

wave-particle duality: In quantum mechanics, a subatomic particle can exist in more than one place at any point in time and can simultaneously behave both as a single particle and as a wave field. This dual state of existence is maintained until the time when the particle's properties are measured, at which point the particle can only be in one place. *See also* quantum uncertainty.

NOTES

Chapter 1. Earth

1. O'Neill, C., Marchi, S., Zhang, S. & Bottke, W. Impact-driven subduction on the Hadean Earth. *Nature Geoscience* 10, 793–97 (2017).

2. O'Neill, C., Lenardic, A., Weller, M., Moresi, L., Quenette, S. & Zhang, S. A window for plate tectonics in terrestrial planet evolution? *Physics of the Earth and Planetary Interiors* 255, 80–92 (2016).

3. Stern, T., Henrys, S. A., Okaya, D., Louie, J. N., Savage, M. K., Lamb, S., Sato, H., Sutherland, R. & Iwasaki, T. A seismic reflection image for the base of a tectonic plate. *Nature* 518, 85–88 (2015).

4. Wan, B., Yang, X., Tian, X., Yuan, H., Kirscher, U. & Mitchell, R. N. Seismological evidence for the earliest global subduction network at 2 Ga ago. *Science Advances* 6, eabc5491 (2020).

5. Tusch, J., Münker, C., Hasenstab, E., Jansen, M., Marien, C. S., Kurzweil, F., Van Kranendonk, M. J., Smithies, H., Maier, W. & Garbe-Schönberg, D. Convective isolation of Hadean mantle reservoirs through Archean time. *Proceedings of the National Academy of Sciences* 118 (2021).

6. Tsekhmistrenko, M., Sigloch, K., Hosseini, K. & Barruol, G. A tree of Indo-African mantle plumes imaged by seismic tomography. *Nature Geoscience* 14, 612–19 (2021).

7. Sayers, D. L. *The Divine Comedy 1: Hell* (Penguin, 1968).

8. Tang, C., Webb, A., Moore, W., Wang, Y., Ma, T. & Chen, T. Breaking Earth's shell into a global plate network. *Nature Communications* 11, 1–6 (2020).

9. Pearson, D. G., Scott, J. M., Liu, J., Schaeffer, A., Wang, L. H., Hunen, J. van, Szilas, K., Chacko, T. & Kelemen, P. B. Deep continental roots and cratons. *Nature* 596, 199–210 (2021).

10. Seeley, J. T. & Wordsworth, R. D. Episodic deluges in simulated hothouse climates. *Nature* 599, 74–79 (2021).

11. Sobolev, S. V. & Brown, M. Surface erosion events controlled the evolution of plate tectonics on Earth. *Nature* 570, 52–57 (2019).

12. Zhao, G., Cawood, P. A., Wilde, S. A. & Sun, M. Review of global 2.1–1.8 Ga orogens: Implications for a pre-Rodinia supercontinent. *Earth-Science Reviews* 59, 125–62 (2002).

13. Cooper, G. F., Macpherson, C. G., Blundy, J. D., Maunder, B., Allen, R. W., Goes, S., Collier, J. S., Bie, L., Harmon, N., Hicks, S. P. & others. Variable water input controls evolution of the Lesser Antilles volcanic arc. *Nature* 582, 525–29 (2020).

14. Jenkins, A., Biggs, J., Rust, A. C. & Rougier, J. C. Decadal timescale correlations between global earthquake activity and volcanic eruption rates. *Geophysical Research Letters* 48, e2021GL093550 (2021).

15. Cornford, F. M. *The Republic of Plato* (London: Oxford University Press, 1966).

16. Green, J., Molloy, J., Davies, H. & Duarte, J. Is there a tectonically driven supertidal cycle? *Geophysical Research Letters* 45, 3568–76 (2018).

17. Rampino, M. R., Caldeira, K. & Zhu, Y. A pulse of the Earth: A 27.5-Myr underlying cycle in coordinated geological events over the last 260 Myr. *Geoscience Frontiers* 12, 101245 (2021).

18. Rampino, M. R., Caldeira, K. & Zhu, Y. A pulse of the Earth: A 27.5-Myr underlying cycle in coordinated geological events over the last 260 Myr. *Geoscience Frontiers* 12, 101245 (2021).

19. Satow, C., Gudmundsson, A., Gertisser, R., Ramsey, C. B., Bazargan, M., Pyle, D. M., Wulf, S., Miles, A. J. & Hardiman, M. Eruptive activity of the Santorini Volcano controlled by sea-level rise and fall. *Nature Geoscience* 14, 586–92 (2021).

20. Hess, B. L., Piazolo, S. & Harvey, J. Lightning strikes as a major facilitator of prebiotic phosphorus reduction on early Earth. *Nature Communications* 12, 1–8 (2021).

21. Klatt, J. M., Chennu, A., Arbic, B. K., Biddanda, B. & Dick, G. J. Possible link between Earth's rotation rate and oxygenation. *Nature Geoscience* 14, 564–70 (2021).

22. Tang, H. & Chen, Y. Global glaciations and atmospheric change at ca. 2.3 ga. *Geoscience Frontiers* 4, 583–96 (2013).

Chapter 2. Life

1. Dembitzer, J., Barkai, R., Ben-Dor, M. & Meiri, S. Levantine overkill: 1.5 million years of hunting down the body size distribution. *Quaternary Science Reviews* 276, 107316 (2022).

2. Roebroeks, W., MacDonald, K., Scherjon, F., Bakels, C., Kindler, L., Nikulina, A., Pop, E. & Gaudzinski-Windheuser, S. Landscape modification by last interglacial Neanderthals. *Science Advances* 7, eabj5567 (2021).

3. Bar-On, Y. M., Phillips, R. & Milo, R. The biomass distribution on earth. *Proceedings of the National Academy of Sciences* 115, 6506–11 (2018).

4. Bradshaw, C. J., Ehrlich, P. R., Beattie, A., Ceballos, G., Crist, E., Diamond, J., Dirzo, R., Ehrlich, A. H., Harte, J., Harte, M. E. & others. Underestimating the challenges of avoiding a ghastly future. *Frontiers in Conservation Science* 1, 1–10 (2021).

5. Moses, M. E. & Brown, J. H. Allometry of human fertility and energy use. *Ecology Letters* 6, 295–300 (2003).

6. Hatton, I. A., Heneghan, R. F., Bar-On, Y. M. & Galbraith, E. D. The global ocean size spectrum from bacteria to whales. *Science Advances* 7, eabh3732 (2021).

7. Womack, T., Crampton, J., Hannah, M. & Collins, K. A positive relationship between functional redundancy and temperature in cenozoic marine ecosystems. *Science* 373, 1027–29 (2021).

8. Kerkhoff, A. J. & Enquist, B. J. The implications of scaling approaches for understanding resilience and reorganization in ecosystems. *Bioscience* 57, 489–99 (2007).

9. Hatton, I. A., Heneghan, R. F., Bar-On, Y. M. & Galbraith, E. D. The global ocean size spectrum from bacteria to whales. *Science Advances* 7, eabh3732 (2021).

10. Hedges, S. B., Marin, J., Suleski, M., Paymer, M. & Kumar, S. Tree of life reveals clock-like speciation and diversification. *Molecular Biology and Evolution* 32, 835–45 (2015).

11. Capra, F. *The Hidden Connections: A Science for Sustainable Living* (Anchor, 2004).

12. Darwin, C. R. *The Origin of Species by Means of Natural Selection*, 286 (London: John Murray, 1876).

13. Wallace, R. A new formal perspective on "Cambrian explosions." *Comptes Rendus Biologies* 337, 1–5 (2014).

14. Wallace, R. A new formal perspective on "Cambrian explosions." *Comptes Rendus Biologies* 337, 1–5 (2014).

15. Hoyal Cuthill, J. F., Guttenberg, N. & Budd, G. E. Impacts of speciation and extinction measured by an evolutionary decay clock. *Nature* 588, 636–41 (2020).

16. Hoyal Cuthill, J. F., Guttenberg, N. & Budd, G. E. Impacts of speciation and extinction measured by an evolutionary decay clock. *Nature* 588, 636–41 (2020).

17. Hoyal Cuthill, J. F., Guttenberg, N. & Budd, G. E. Impacts of speciation and extinction measured by an evolutionary decay clock. *Nature* 588, 636–41 (2020).

18. Crampton, J. S., Cooper, R. A., Foote, M. & Sadler, P. M. Ephemeral species in the fossil record? Synchronous coupling of macroevolutionary dynamics in mid-Paleozoic zooplankton. *Paleobiology* 46, 123–35 (2020).

19. Hoyal Cuthill, J. F., Guttenberg, N. & Budd, G. E. Impacts of speciation and extinction measured by an evolutionary decay clock. *Nature* 588, 636–41 (2020).

20. Kirchner, J. W. & Weil, A. Delayed biological recovery from extinctions throughout the fossil record. *Nature* 404, 177–80 (2000).

21. Burgess, S. D., Bowring, S. & Shen, S.-Z. High-precision timeline for earth's most severe extinction. *Proceedings of the National Academy of Sciences* 111, 3316–21 (2014).

22. Grasby, S. E., Sanei, H. & Beauchamp, B. Catastrophic dispersion of coal fly ash into oceans during the latest Permian extinction. *Nature Geoscience* 4, 104–7 (2011).

23. Smith, A. *The Theory of Moral Sentiments* (Dover Publications, 1759).

24. Daver, G., Guy, F., Mackaye, H., Likius, A., Boisserie, J.-R., Moussa, A., Pallas, L., Vignaud, P. & Clarisse, N. Postcranial evidence of late Miocene hominin bipedalism in Chad. *Nature* 1–7 (2022).

25. Kidwell, S. M. Biology in the Anthropocene: Challenges and insights from young fossil records. *Proceedings of the National Academy of Sciences* 112, 4922–29 (2015).

26. Kidwell, S. M. Biology in the Anthropocene: Challenges and insights from young fossil records. *Proceedings of the National Academy of Sciences* 112, 4922–29 (2015).

27. Bascompte, J., García, M. B., Ortega, R., Rezende, E. L. & Pironon, S. Mutualistic interactions reshuffle the effects of climate change on plants across the tree of life. *Science Advances* 5, eaav2539 (2019).

28. Fricke, E. C., Ordonez, A., Rogers, H. S. & Svenning, J.-C. The effects of defaunation on plants' capacity to track climate change. *Science* 375, 210–14 (2022).

29. Campbell-Staton, S. C., Arnold, B. J., Gonçalves, D., Granli, P., Poole, J., Long, R. A. & Pringle, R. M. Ivory poaching and the rapid evolution of tusklessness in African elephants. *Science* 374, 483–87 (2021).

30. Zaveri, E., Russ, J. & Damania, R. Rainfall anomalies are a significant driver of cropland expansion. *Proceedings of the National Academy of Sciences* 117, 10225–33 (2020).

31. Mrad, A., Katul, G. G., Levia, D. F., Guswa, A. J., Boyer, E. W., Bruen, M., Carlyle-Moses, D. E., Coyte, R., Creed, I. F., Van De Giesen, N. & others. Peak grain forecasts for the US High Plains amid withering waters. *Proceedings of the National Academy of Sciences* 117, 26145–50 (2020).

32. Florido Ngu, F., Kelman, I., Chambers, J. & Ayeb-Karlsson, S. Correlating heatwaves and relative humidity with suicide (fatal intentional self-harm). *Scientific Reports* 11, 1–9 (2021).

33. Gardner, C. J. & Wordley, C. F. Scientists must act on our own warnings to humanity. *Nature Ecology & Evolution* 3, 1271–72 (2019).

34. Trisos, C. H., Merow, C. & Pigot, A. L. The projected timing of abrupt ecological disruption from climate change. *Nature* 580, 496–501 (2020).

35. Colwell, R. K. Spatial scale and the synchrony of ecological disruption. *Nature* 599, E8–E10 (2021).

36. Bush, E. R., Whytock, R. C., Bahaa-El-Din, L., Bourgeois, S., Bunnefeld, N., Cardoso, A. W., Dikangadissi, J. T., Dimbonda, P., Dimoto, E., Edzang Ndong, J. & others. Long-term collapse in fruit availability threatens Central African forest megafauna. *Science* 370, 1219–22 (2020).

37. Leung, B., Hargreaves, A. L., Greenberg, D. A., McGill, B., Dornelas, M. & Freeman, R. Clustered versus catastrophic global vertebrate declines. *Nature* 588, 267–71 (2020).

38. Leclère, D., Obersteiner, M., Barrett, M., Butchart, S. H., Chaudhary, A., De Palma, A., DeClerck, F. A., Di Marco, M., Doelman, J. C., Dürauer, M. & others. Bending the curve of terrestrial biodiversity needs an integrated strategy. *Nature* 585, 551–56 (2020).

39. Gibb, R., Redding, D. W., Chin, K. Q., Donnelly, C. A., Blackburn, T. M., Newbold, T. & Jones, K. E. Zoonotic host diversity increases in human-dominated ecosystems. *Nature* 584, 398–402 (2020).

40. Day, T., Kennedy, D. A., Read, A. F. & McAdams, D. The economics of managing evolution. *PLoS Biology* 19, e3001409 (2021).

41. Lehman, C., Loberg, S., Wilson, M. & Gorham, E. Ecology of the Anthropocene signals hope for consciously managing the planetary ecosystem. *Proceedings of the National Academy of Sciences* 118 (2021).

42. Knoll, A. H. & Nowak, M. A. The timetable of evolution. *Science Advances* 3, e1603076 (2017).

Chapter 3. Climate

1. Foote, E. Circumstances affecting the heat of the sun's rays. *American Journal of Science and Arts* 22, no. 66, 382–83 (November 1856).

2. Callendar, G. S. The artificial production of carbon dioxide and its influence on temperature. *Quarterly Journal of the Royal Meteorological Society* 64, 223–40 (1938).

3. Reck, Ruth A. Aerosols and polar temperature changes. *Science* 188, 728–30 (1975).

4. Mercer, J. H. West Antarctic ice sheet and CO2 greenhouse effect: A threat of disaster. *Nature* 271, 321–25 (1978).

5. Scott, A. C. & Glasspool, I. J. The diversification of Paleozoic fire systems and fluctuations in atmospheric oxygen concentration. *Proceedings of the National Academy of Sciences* 103, 10861–65 (2006).

6. Dudley, R. Atmospheric oxygen, giant Paleozoic insects and the evolution of aerial locomotor performance. *The Journal of Experimental Biology* 201, 1043–50 (1998).

7. Ganopolski, A., Winkelmann, R. & Schellnhuber, H. J. Critical insolation–CO_2 relation for diagnosing past and future glacial inception. *Nature* 529, 200–203 (2016).

8. Hawkins, E., Ortega, P., Suckling, E., Schurer, A., Hegerl, G., Jones, P., Joshi, M., Osborn, T. J., Masson-Delmotte, V., Mignot, J. & others. Estimating changes in global temperature since the preindustrial period. *Bulletin of the American Meteorological Society* 98, 1841–56 (2017).

9. Abram, N. J., McGregor, H. V., Tierney, J. E., Evans, M. N., McKay, N. P., Kaufman, D. S., Thirumalai, K., Martrat, B., Goosse, H., Phipps, S. J. & others. Early onset of industrial-era warming across the oceans and continents. *Nature* 536, 411 (2016).

10. Osman, M. B., Tierney, J. E., Zhu, J., Tardif, R., Hakim, G. J., King, J. & Poulsen, C. J. Globally resolved surface temperatures since the last glacial maximum. *Nature* 599, 239–44 (2021).

11. Lyell, C. *Principles of geology, Being an Attempt to Explain the Former Changes of the Earth's Surface by References to Causes Now in Operation*, vol. 1 (Murray, 1830).

12. Agassiz, L. On glaciers, and the evidence of their having once existed in Scotland, Ireland, and England. *Proceedings of the Geological Society of London* 3, 327–32 (1841).

13. Croll, J. On the physical cause of the change of climate during geological epochs. *Philosophical Magazine* 28, 121–37 (1864).

14. Hays, J. D., Imbrie, J. & Shackleton, N. J. Variations in the earth's orbit; pacemaker of the ice ages. *Science* 194, 1121–132.

15. Hays, J. D., Imbrie, J. & Shackleton, N. J. Variations in the earth's orbit; pacemaker of the ice ages. *Science* 194, 1121–132.

16. Lorenz, E. N. Deterministic nonperiodic flow. *Journal of Atmospheric Sciences* 20, 130–41 (1963).

17. Genthon, C., Barnola, J. M., Raynaud, D., Lorius, C., Jouzel, J., Barkov, N. I., Korotkevich, Y. S. & Kotlyakov, V. M. Vostok ice core: climatic response to CO_2 and orbital forcing changes over the last climatic cycle. *Nature* 329, 414–18 (1987); Tarasov, L. & Peltier, W. R. Terminating the 100 kyr ice age cycle. *Journal of Geophysical Research* 102, 21665–93 (1997).

18. Loutre, M. & Berger, A. No glacial-interglacial cycle in the ice volume simulated under a constant astronomical forcing and a variable CO_2. *Geophysical Research Letters* 27, 783–86; Timmermann, A., Friedrich, T., Timm, O. E., Chikamoto, M., Abe-Ouchi, A. & Ganopolski, A. Modeling obliquity and CO_2 effects on Southern Hemisphere climate during the past 408 ka. *Journal of Climate* 27, 1863–75 (2014).

19. Peltier, W. R. & Marshall, S. Coupled energy-balance/ice-sheet model simulations of the glacial cycle: A possible connection between terminations and terrigenous dust. *Journal of Geophysical Research* 100, 14269–89 (1995).

20. Hyde, W. & Peltier, W. Sensitivity experiments with a model of the ice age cycle: The response to harmonic forcing. *Journal of the Atmospheric Sciences*

42, 2170–88 (1985); Abe-Ouchi, A., Saito, F., Kawamura, K., Raymo, M. E., Okuno, J., Takahashi, K. & Blatter, H. Insolation-driven 100,000-year glacial cycles and hysteresis of ice-sheet volume. *Nature* 500, 190–94 (2013).

21. Clark, P. U. & Pollard, D. Origin of the middle Pleistocene transition by ice sheet erosion of regolith. *Paleoceanography* 13, 1–9 (1998); Willeit, M., Ganopolski, A., Calov, R. & Brovkin, V. Mid-Pleistocene transition in glacial cycles explained by declining CO_2 and regolith removal. *Science Advances* 5, eaav7337 (2019).

22. Gildor, H. & Tziperman, E. Sea ice as the glacial cycles' climate switch: Role of seasonal and orbital forcing. *Paleoceanography* 15, 605–15 (2000); Lee, J.-E., Shen, A., Fox-Kemper, B. & Ming, Y. Hemispheric sea ice distribution sets the glacial tempo. *Geophysical Research Letters* 44, 1008–14 (2017).

23. Zachos, J., Pagani, M., Sloan, L., Thomas, E. & Billups, K. Trends, rhythms, and aberrations in global climate 65 Ma to present. *Science* 292, 686–93 (2001).

24. Lisiecki, L. E. & Raymo, M. E. A Pliocene-Pleistocene stack of 57 globally distributed benthic $\delta18O$ records. *Paleoceanography* 20, PA1003 (2005).

25. Alley, R. B. The younger dryas cold interval as viewed from central Greenland. *Quaternary Science Reviews* 19, 213–26 (2000).

26. Watts, A. W. *The Way of Zen* (Penguin, 1980).

27. Huybers, P. & Wunsch, C. Obliquity pacing of the late Pleistocene glacial terminations. *Nature* 424, 491–94 (2005).

28. Pelletier, J. D. Coherence resonance and ice ages. *Journal of Geophysical Research* 108, D20 (2003).

29. Maslov, L. A. Self-organization of the Earth's climate system versus Milankovitch–Berger astronomical cycles. *Journal of Advances in Modeling Earth Systems* 6, 650–57 (2014). Mitsui, T., Crucifix, M. & Aihara, K. Bifurcations and strange nonchaotic attractors in a phase oscillator model of glacial–interglacial cycles. *Physica D: Nonlinear Phenomena* 306, 25–33 (2015).

30. Verhulst, P.-F. Recherches mathématiques sur la loi d'accroissement de la population. *Nouveaux Mémoires de l'Académie Royale des Sciences et Belles-Lettres de Bruxelles* 18, 1–38 (1845).

31. May, R. M. Limit cycles in predator-prey communities. *Science* 177, 900–902 (1972).

32. Steffen, W., Rockström, J., Richardson, K., Lenton, T. M., Folke, C., Liverman, D., Summerhayes, C. P., Barnosky, A. D., Cornell, S. E., Crucifix, M. & others. Trajectories of the Earth system in the Anthropocene. *Proceedings of the National Academy of Sciences* 115, 8252–59 (2018).

33. Bradshaw, C. J., Ehrlich, P. R., Beattie, A., Ceballos, G., Crist, E., Diamond, J., Dirzo, R., Ehrlich, A. H., Harte, J., Harte, M. E. & others. Underestimating the challenges of avoiding a ghastly future. *Frontiers in Conservation Science* 1, 1–10 (2021).

34. Feulner, G. Formation of most of our coal brought Earth close to global glaciation. *Proceedings of the National Academy of Sciences* 114, 11333–37 (2017).

Chapter 4. Humans

1. Collins, M. *Carrying the Fire: An Astronaut's Journeys* (New York: Cooper Square Press, 2001).

2. Nietzsche, F. *The Birth of Tragedy out of the Spirit of Music*, trans. S. Whiteside, ed. Michael Tanner (London: Penguin, 1993).

3. Almécija, S., Hammond, A. S., Thompson, N. E., Pugh, K. D., Moyà-Solà, S. & Alba, D. M. Fossil apes and human evolution. *Science* 372, eabb4363 (2021).

4. Leakey, M. D. & Hay, R. L. Pliocene footprints in the Laetolil Beds at Laetoli, northern Tanzania. *Nature* 278, 317–23 (1979).

5. McNutt, E. J., Hatala, K. G., Miller, C., Adams, J., Casana, J., Deane, A. S., Dominy, N. J., Fabian, K., Fannin, L. D., Gaughan, S. & others. Footprint evidence of early hominin locomotor diversity at Laetoli, Tanzania. *Nature* 600, 468–71 (2021).

6. Ben-Dor, M. & Barkai, R. Prey size decline as a unifying ecological selecting agent in Pleistocene human evolution. *Quaternary* 4, 7 (2021).

7. Chan, E. K., Timmermann, A., Baldi, B. F., Moore, A. E., Lyons, R. J., Lee, S.-S., Kalsbeek, A. M., Petersen, D. C., Rautenbach, H., Förtsch, H. E. & others. Human origins in a southern African palaeo-wetland and first migrations. *Nature* 575, 185–89 (2019).

8. Beyer, R. M., Krapp, M., Eriksson, A. & Manica, A. Climatic windows for human migration out of Africa in the past 300,000 years. *Nature Communications* 12, 1–10 (2021).

9. Will, M., Krapp, M., Stock, J. T. & Manica, A. Different environmental variables predict body and brain size evolution in *Homo*. *Nature Communications* 12, 1–12 (2021).

10. Parikh, R., Sorek, E., Parikh, S., Michael, K., Bikovski, L., Tshori, S., Shefer, G., Mingelgreen, S., Zornitzki, T., Knobler, H. & others. Skin exposure to UVB light induces a skin-brain-gonad axis and sexual behavior. *Cell Reports* 36, 109579 (2021).

11. MacDonald, K., Scherjon, F., Veen, E. van, Vaesen, K. & Roebroeks, W. Middle Pleistocene fire use: The first signal of widespread cultural diffusion in human evolution. *Proceedings of the National Academy of Sciences* 118, e2101108118 (2021).

12. MacDonald, K., Scherjon, F., Veen, E. van, Vaesen, K. & Roebroeks, W. Middle Pleistocene fire use: The first signal of widespread cultural diffusion in human evolution. *Proceedings of the National Academy of Sciences* 118, e2101108118 (2021).

13. Hallett, E. Y., Marean, C. W., Steele, T. E., Álvarez-Fernández, E., Jacobs, Z., Cerasoni, J. N., Aldeias, V., Scerri, E. M., Olszewski, D. I., El Hajraoui, M. A. & others. A worked bone assemblage from 120,000–90,000 year old deposits at Contrebandiers Cave, Atlantic Coast, Morocco. *Iscience* 24, 102988 (2021).

14. Mackay, A., Armitage, S. J., Niespolo, E. M., Sharp, W. D., Stahlschmidt, M. C., Blackwood, A. F., Boyd, K. C., Chase, B. M., Lagle, S. E., Kaplan,

C. F. & others. Environmental influences on human innovation and behavioral diversity in southern Africa 92–80 thousand years ago. *Nature Ecology & Evolution* 6, 361–69 (2022).

15. Coutrot, A., Manley, E., Goodroe, S., Gahnstrom, C., Filomena, G., Yesiltepe, D., Dalton, R., Wiener, J. M., Hölscher, C., Hornberger, M. & others. Entropy of city street networks linked to future spatial navigation ability. *Nature* 604, 104–10 (2022).

16. Li, W., Wang, Z., Syed, S., Lyu, C., Lincoln, S., O'Neil, J., Nguyen, A. D., Feng, I. & Young, M. W. Chronic social isolation signals starvation and reduces sleep in Drosophila. *Nature* 597, 239–44 (2021).

17. Thomas, A. J., Woo, B., Nettle, D., Spelke, E. & Saxe, R. Early concepts of intimacy: Young humans use saliva sharing to infer close relationships. *Science* 375, 311–15 (2022).

18. Hu, Y., McElwain, N. L. & Berry, D. Mother-child mutually responsive orientation and real-time physiological coordination. *Developmental Psychobiology* 63, e22200 (2021).

19. Fernandez, A. A., Burchardt, L. S., Nagy, M. & Knörnschild, M. Babbling in a vocal learning bat resembles human infant babbling. *Science* 373, 923–26 (2021).

20. Thibault, S., Py, R., Gervasi, A. M., Salemme, R., Koun, E., Lövden, M., Boulenger, V., Roy, A. C. & Brozzoli, C. Tool use and language share syntactic processes and neural patterns in the basal ganglia. *Science* 374, eabe0874 (2021).

21. Heilbron, M., Armeni, K., Schoffelen, J.-M., Hagoort, P. & De Lange, F. P. A hierarchy of linguistic predictions during natural language comprehension. *Proceedings of the National Academy of Sciences* 119, e2201968119 (2022).

22. Hofstetter, S., Zuiderbaan, W., Heimler, B., Dumoulin, S. O. & Amedi, A. Topographic maps and neural tuning for sensory substitution dimensions learned in adulthood in a congenital blind subject. *NeuroImage* 235, 118029 (2021).

23. Twomey, C. R., Roberts, G., Brainard, D. H. & Plotkin, J. B. What we talk about when we talk about colors. *Proceedings of the National Academy of Sciences* 118, e2109237118 (2021).

24. Krumpholz, C., Quigley, C., Ameen, K., Reuter, C., Fusani, L. & Leder, H. The effects of pitch manipulation on male ratings of female speakers and their voices. *Frontiers in Psychology* 13 (2022).

25. Norman-Haignere, S. V., Feather, J., Boebinger, D., Brunner, P., Ritaccio, A., McDermott, J. H., Schalk, G. & Kanwisher, N. A neural population selective for song in human auditory cortex. *Current Biology* 32, 1470–84 (2022).

26. Chatwin, B. *The Songlines* (London: Picador, 1988), 325.

27. Pontzer, H., Yamada, Y., Sagayama, H., Ainslie, P. N., Andersen, L. F., Anderson, L. J., Arab, L., Baddou, I., Bedu-Addo, K., Blaak, E. E. & others. Daily energy expenditure through the human life course. *Science* 373, 808–12 (2021).

28. Bramble, D. M. & Lieberman, D. E. Endurance running and the evolution of Homo. *Nature* 432, 345–52 (2004).

29. Barr, W. A., Pobiner, B., Rowan, J., Du, A. & Faith, J. T. No sustained increase in zooarchaeological evidence for carnivory after the appearance of *Homo erectus*. *Proceedings of the National Academy of Sciences* 119, e2115540119 (2022).

30. Kraft, T. S., Venkataraman, V. V., Wallace, I. J., Crittenden, A. N., Holowka, N. B., Stieglitz, J., Harris, J., Raichlen, D. A., Wood, B., Gurven, M. & others. The energetics of uniquely human subsistence strategies. *Science* 374, eabf0130 (2021).

31. Blum, W. E., Zechmeister-Boltenstern, S. & Keiblinger, K. M. Does soil contribute to the human gut microbiome? *Microorganisms* 7, 287 (2019).

32. Mayer, E. *The Gut-Immune Connection: How Understanding the Connection between Food and Immunity Can Help Us Regain Our Health* (New York: HarperCollins, 2021).

33. Heys, C., Fisher, A., Dewhurst, A., Lewis, Z. & Lizé, A. Exposure to foreign gut microbiota can facilitate rapid dietary shifts. *Scientific Reports* 11, 1–6 (2021); Letourneau, J., Holmes, Z. C., Dallow, E. P., Durand, H. K., Jiang, S., Carrion, V. M., Gupta, S. K., Mincey, A. C., Muehlbauer, M. J., Bain, J. R. & others. Ecological memory of prior nutrient exposure in the human gut microbiome. *The ISME Journal* 1–12 (2022).

34. Liebenberg, L. *The Art of Tracking: The Origin of Science* (Claremont, South Africa: David Phillip Publishers, 1990), 176.

35. Quinn, D. *Ishmael* (New York: Bantam, 1995), 263.

36. McFarland, M. J., Hauer, M. E. & Reuben, A. Half of US population exposed to adverse lead levels in early childhood. *Proceedings of the National Academy of Sciences* 119, e2118631119 (2022).

37. Long, Y., Chen, Q., Larsson, H. & Rzhetsky, A. Observable variations in human sex ratio at birth. *PLoS Computational Biology* 17, e1009586 (2021).

38. Wolter, M., Grant, E. T., Boudaud, M., Steimle, A., Pereira, G. V., Martens, E. C. & Desai, M. S. Leveraging diet to engineer the gut microbiome. *Nature Reviews Gastroenterology & Hepatology* 18, 885–902 (2021).

39. Swift, C. L., Louie, K. B., Bowen, B. P., Olson, H. M., Purvine, S. O., Salamov, A., Mondo, S. J., Solomon, K. V., Wright, A. T., Northen, T. R. & others. Anaerobic gut fungi are an untapped reservoir of natural products. *Proceedings of the National Academy of Sciences* 118, e2019855118 (2021).

40. Stulp, G., Barrett, L., Tropf, F. C. & Mills, M. Does natural selection favor taller stature among the tallest people on earth? *Proceedings of the Royal Society B: Biological Sciences* 282, 20150211 (2015).

41. Longrich, N. R. Future evolution: From looks to brains and personality, how will humans change in the next 10,000 years? *The Conversation* (2022). https://theconversation.com/future-evolution-from-looks-to-brains-and-personality-how-will-humans-change-in-the-next-10-000-years-176997.

42. Melamed, D., Nov, Y., Malik, A., Yakass, M. B., Bolotin, E., Shemer, R., Hiadzi, E. K., Skorecki, K. L. & Livnat, A. De novo mutation rates at the single-mutation resolution in a human HBB gene region associated with adaptation and genetic disease. *Genome Research* 32, 488–98 (2022).

43. Andermann, T., Faurby, S., Turvey, S. T., Antonelli, A. & Silvestro, D. The past and future human impact on mammalian diversity. *Science Advances* 6, eabb2313 (2020).

44. Bologna, M. & Aquino, G. Deforestation and world population sustainability: A quantitative analysis. *Scientific Reports* 10, 1–9 (2020).

45. Jiang, J. H., Feng, F., Rosen, P. E., Fahy, K. A., Das, P., Obacz, P., Zhang, A. & Zhu, Z.-H. Avoiding the Great Filter: Predicting the timeline for humanity to reach Kardashev Type I civilization. *Galaxies* 10, 68 (2022).

46. Vanchurin, V., Wolf, Y. I., Koonin, E. V. & Katsnelson, M. I. Thermodynamics of evolution and the origin of life. *Proceedings of the National Academy of Sciences* 119, e2120042119 (2022).

47. Vanchurin, V., Wolf, Y. I., Katsnelson, M. I. & Koonin, E. V. Toward a theory of evolution as multilevel learning. *Proceedings of the National Academy of Sciences* 119, e2120037119 (2022).

48. Papers of John F. Kennedy. Presidential Papers. Address at Rice University, Houston, Texas, 12 September 1962. www.jfklibrary.org/asset-viewer/archives/ JFKPOF/040/JFKPOF-040-001.

Chapter 5. Society

1. Stewart, H., Bradwell, T., Bullard, J., Davies, S., Golledge, N. & McCulloch, R. 8000 years of North Atlantic storminess reconstructed from a Scottish peat record: Implications for Holocene atmospheric circulation patterns in Western Europe. *Journal of Quaternary Science* 32, 1075–84 (2017).

2. Pearson, M. P., Pollard, J., Richards, C., Welham, K., Kinnaird, T., Shaw, D., Simmons, E., Stanford, A., Bevins, R., Ixer, R. & others. The original Stonehenge? A dismantled stone circle in the Preseli Hills of west Wales. *Antiquity* 95, 85–103 (2021).

3. Patterson, N., Isakov, M., Booth, T., Büster, L., Fischer, C.-E., Olalde, I., Ringbauer, H., Akbari, A., Cheronet, O., Bleasdale, M. & others. Large-scale migration into Britain during the middle to late Bronze Age. *Nature* 601, 588–94 (2022).

4. Crabtree, S. A., White, D. A., Bradshaw, C. J., Saltré, F., Williams, A. N., Beaman, R. J., Bird, M. I. & Ulm, S. Landscape rules predict optimal superhighways for the first peopling of Sahul. *Nature Human Behavior* 5, 1303–13 (2021).

5. Miller, J. M. & Wang, Y. V. Ostrich eggshell beads reveal 50,000-year-old social network in Africa. *Nature* 601, 234–39 (2022).

6. Harari, Y. N. *Sapiens: A Brief History of Humankind* (New York: Random House, 2014).

7. Amit, R., Enzel, Y. & Crouvi, O. Quaternary influx of proximal coarse-grained dust altered circum-Mediterranean soil productivity and impacted early human culture. *Geology* 49, 61–65 (2021).

8. McConnell, J. R., Sigl, M., Plunkett, G., Burke, A., Kim, W. M., Raible, C. C., Wilson, A. I., Manning, J. G., Ludlow, F., Chellman, N. J. & others.

Extreme climate after massive eruption of Alaska's Okmok volcano in 43 BCE and effects on the late Roman Republic and Ptolemaic kingdom. *Proceedings of the National Academy of Sciences* 117, 15443–49 (2020).

9. Strunz, S. & Braeckel, O. Did volcano eruptions alter the trajectories of the Roman Republic and the Ptolemaic kingdom? Moving beyond black-box determinism. *Proceedings of the National Academy of Sciences* 117, 32207–8 (2020).

10. Huhtamaa, H., Stoffel, M. & Corona, C. Recession or resilience? Long-range socioeconomic consequences of the 17th century volcanic eruptions in the far north. *Climate of the Past* 18, 2077–92 (2022).

11. Domínguez-Andrés, J., Kuijpers, Y., Bakker, O. B., Jaeger, M., Xu, C.-J., Van der Meer, J. W., Jakobsson, M., Bertranpetit, J., Joosten, L. A., Li, Y. & others. Evolution of cytokine production capacity in ancient and modern European populations. *Elife* 10, e64971 (2021).

12. Kennett, D. J., Masson, M., Lope, C. P., Serafin, S., George, R. J., Spencer, T. C., Hoggarth, J. A., Culleton, B. J., Harper, T. K., Prufer, K. M. & others. Drought-induced civil conflict among the ancient Maya. *Nature Communications* 13, 1–10 (2022).

13. Jun, T. & Sethi, R. Extreme weather events and military conflict over seven centuries in ancient Korea. *Proceedings of the National Academy of Sciences* 118, e2021976118 (2021).

14. Zhang, D. D., Brecke, P., Lee, H. F., He, Y.-Q. & Zhang, J. Global climate change, war, and population decline in recent human history. *Proceedings of the National Academy of Sciences* 104, 19214–19 (2007).

15. Degroot, D., Anchukaitis, K., Bauch, M., Burnham, J., Carnegy, F., Cui, J., Luna, K. de, Guzowski, P., Hambrecht, G., Huhtamaa, H. & others. Towards a rigorous understanding of societal responses to climate change. *Nature* 591, 539–50 (2021).

16. Foer, J. S. *We Are the Weather: Saving the Planet Begins at Breakfast* (London: Penguin UK, 2019).

17. Yabe, T., Rao, P. S. C., Ukkusuri, S. V. & Cutter, S. L. Toward data-driven, dynamical complex systems approaches to disaster resilience. *Proceedings of the National Academy of Sciences* 119, e2111997119 (2022).

18. Acosta, R. J., Kishore, N., Irizarry, R. A. & Buckee, C. O. Quantifying the dynamics of migration after Hurricane Maria in Puerto Rico. *Proceedings of the National Academy of Sciences* 117, 32772–78 (2020).

19. Nix-Stevenson, D. Human response to natural disasters. *Sage Open* 3, 2158244013489684 (2013).

20. Preston, J., Chadderton, C., Kitagawa, K. & Edmonds, C. Community response in disasters: An ecological learning framework. *International Journal of Lifelong Education* 34, 727–53 (2015).

21. Barceló, J. The long-term effects of war exposure on civic engagement. *Proceedings of the National Academy of Sciences* 118, e2015539118 (2021).

22. Drelichman, M., Vidal-Robert, J. & Voth, H.-J. The long-run effects of religious persecution: Evidence from the Spanish Inquisition. *Proceedings of the National Academy of Sciences* 118, e2022881118 (2021).

23. Bai, X., Ramos, M. R. & Fiske, S. T. As diversity increases, people paradoxically perceive social groups as more similar. *Proceedings of the National Academy of Sciences* 117, 12741–49 (2020).

24. Atherton, G., Sebanz, N. & Cross, L. Imagine all the synchrony: The effects of actual and imagined synchronous walking on attitudes towards marginalised groups. *PloS One* 14, e0216585 (2019).

25. Bollen, J., Ten Thij, M., Breithaupt, F., Barron, A. T., Rutter, L. A., Lorenzo-Luaces, L. & Scheffer, M. Historical language records reveal a surge of cognitive distortions in recent decades. *Proceedings of the National Academy of Sciences* 118, e2102061118 (2021).

26. Brumm, H., Goymann, W., Derégnaucourt, S., Geberzahn, N. & Zollinger, S. A. Traffic noise disrupts vocal development and suppresses immune function. *Science Advances* 7, eabe2405 (2021).

27. Giddens, N. T., Juneau, P., Manza, P., Wiers, C. E. & Volkow, N. D. Disparities in sleep duration among American children: Effects of race and ethnicity, income, age, and sex. *Proceedings of the National Academy of Sciences* 119, e2120009119 (2022).

28. Simon, E. B., Vallat, R., Rossi, A. & Walker, M. P. Sleep loss leads to the withdrawal of human helping across individuals, groups, and large-scale societies. *PLoS Biology* 20, e3001733 (2022).

29. Testard, C., Brent, L. J., Andersson, J., Chiou, K. L., Negron-Del Valle, J. E., DeCasien, A. R., Acevedo-Ithier, A., Stock, M. K., Antón, S. C., Gonzalez, O. & others. Social connections predict brain structure in a multidimensional free-ranging primate society. *Science Advances* 8, eabl5794 (2022).

30. Riedl, C., Kim, Y. J., Gupta, P., Malone, T. W. & Woolley, A. W. Quantifying collective intelligence in human groups. *Proceedings of the National Academy of Sciences* 118, e2005737118 (2021).

31. Almaatouq, A., Alsobay, M., Yin, M. & Watts, D. J. Task complexity moderates group synergy. *Proceedings of the National Academy of Sciences* 118, e2101062118 (2021).

32. Yang, V. C., Galesic, M., McGuinness, H. & Harutyunyan, A. Dynamical system model predicts when social learners impair collective performance. *Proceedings of the National Academy of Sciences* 118, e2106292118 (2021).

33. Bak-Coleman, J. B., Alfano, M., Barfuss, W., Bergstrom, C. T., Centeno, M. A., Couzin, I. D., Donges, J. F., Galesic, M., Gersick, A. S., Jacquet, J. & others. Stewardship of global collective behavior. *Proceedings of the National Academy of Sciences* 118, e2025764118 (2021).

34. Chetty, R., Jackson, M. O., Kuchler, T., Stroebel, J., Hendren, N., Fluegge, R. B., Gong, S., Gonzalez, F., Grondin, A., Jacob, M. & others. Social capital I: Measurement and associations with economic mobility. *Nature* 608, 108–

21 (2022); Chetty, R., Jackson, M. O., Kuchler, T., Stroebel, J., Hendren, N., Fluegge, R. B., Gong, S., Gonzalez, F., Grondin, A., Jacob, M. & others. Social capital II: Determinants of economic connectedness. *Nature* 608, 122–34 (2022).

35. Su, Q., McAvoy, A., Mori, Y. & Plotkin, J. B. Evolution of prosocial behaviors in multilayer populations. *Nature Human Behaviour* 6, 338–48 (2022); Su, Q., Allen, B. & Plotkin, J. B. Evolution of cooperation with asymmetric social interactions. *Proceedings of the National Academy of Sciences* 119, e2113468118 (2022).

36. Freud, S. *Civilization and Its Discontents*, trans. David McLintock (London: Penguin, 2002).

37. Van Doesum, N. J., Murphy, R. O., Gallucci, M., Aharonov-Majar, E., Athenstaedt, U., Au, W. T., Bai, L., Böhm, R., Bovina, I., Buchan, N. R. & others. Social mindfulness and prosociality vary across the globe. *Proceedings of the National Academy of Sciences* 118, e2023846118 (2021).

38. Zak, P. J., Curry, B., Owen, T. & Barraza, J. A. Oxytocin release increases with age and is associated with life satisfaction and prosocial behaviors. *Frontiers in Behavioral Neuroscience* 119 (2022).

39. Nanakdewa, K., Madan, S., Savani, K. & Markus, H. R. The salience of choice fuels independence: Implications for self-perception, cognition, and behavior. *Proceedings of the National Academy of Sciences* 118, e2021727118 (2021).

40. Theriault, J. E., Young, L. & Barrett, L. F. The sense of should: A biologically-based framework for modeling social pressure. *Physics of Life Reviews* 36, 100–136 (2021).

41. Zampetaki, A. V., Liebchen, B., Ivlev, A. V. & Löwen, H. Collective self-optimization of communicating active particles. *Proceedings of the National Academy of Sciences* 118, e2111142118 (2021).

42. Dunbar, R. I. Neocortex size as a constraint on group size in primates. *Journal of Human Evolution* 22, 469–93 (1992).

43. West, B., Massari, G., Culbreth, G., Failla, R., Bologna, M., Dunbar, R. & Grigolini, P. Relating size and functionality in human social networks through complexity. *Proceedings of the National Academy of Sciences* 117, 18355–58 (2020).

44. Ye, M., Zino, L., Mlakar, Ž., Bolderdijk, J. W., Risselada, H., Fennis, B. M. & Cao, M. Collective patterns of social diffusion are shaped by individual inertia and trend-seeking. *Nature Communications* 12, 1–12 (2021).

45. Dalege, J. & Does, T. van der. Using a cognitive network model of moral and social beliefs to explain belief change. *Science Advances* 8, eabm0137 (2022).

46. Andreoni, J., Nikiforakis, N. & Siegenthaler, S. Predicting social tipping and norm change in controlled experiments. *Proceedings of the National Academy of Sciences* 118, e2014893118 (2021).

47. Mengelkoch, S., Gassen, J., Prokosch, M. L., Boehm, G. W. & Hill, S. E. More than just a pretty face? The relationship between immune function and perceived facial attractiveness. *Proceedings of the Royal Society B* 289, 20212476 (2022).

48. Freud, S. *Civilization and Its Discontents*, trans. David McLintock (London: Penguin, 2002).

49. Henri, R. *The Art Spirit: Notes, Articles, Fragments of Letters and Talks to Students, Bearing on the Concept and Technique of Picture Making, the Study of Art Generally, and on Appreciation* (Boulder, CO: Westview Press, 1984).

50. Aristotle, *Nicomachean Ethics*, ed. Hugh Treddenick (London: Penguin, 2004).

51. Mengelkoch, S., Gassen, J., Prokosch, M. L., Boehm, G. W. & Hill, S. E. More than just a pretty face? The relationship between immune function and perceived facial attractiveness. *Proceedings of the Royal Society B* 289, 20212476 (2022).

52. Hunter, R. L. *Plato's Symposium* (Oxford, UK: Oxford University Press on Demand, 2004).

53. Markow, T. A. "Cost" of virginity in wild *Drosophila melanogaster* females. *Ecology and Evolution* 1, 596–600 (2011).

54. Worthington, A. M. & Kelly, C. D. Females gain survival benefits from immune-boosting ejaculates. *Evolution* 70, 928–33 (2016).

55. Glasper, E. R., LaMarca, E. A., Bocarsly, M. E., Fasolino, M., Opendak, M. & Gould, E. Sexual experience enhances cognitive flexibility and dendritic spine density in the medial prefrontal cortex. *Neurobiology of Learning and Memory* 125, 73–79 (2015); Glasper, E. R. & Gould, E. Sexual experience restores age-related decline in adult neurogenesis and hippocampal function. *Hippocampus* 23, 303–12 (2013).

56. Li, L., Huang, X., Xiao, J., Zheng, Q., Shan, X., He, C., Liao, W., Chen, H., Menon, V. & Duan, X. Neural synchronization predicts marital satisfaction. *Proceedings of the National Academy of Sciences* 119, e2202515119 (2022).

57. Templeton, E. M., Chang, L. J., Reynolds, E. A., Cone LeBeaumont, M. D. & Wheatley, T. Fast response times signal social connection in conversation. *Proceedings of the National Academy of Sciences* 119, e2116915119 (2022).

58. Thomas, J. O. & Dubar, R. T. Disappearing in the age of hypervisibility: Definition, context, and perceived psychological consequences of social media ghosting. *Psychology of Popular Media* 10, 291 (2021).

59. Henri, R. *The Art Spirit: Notes, Articles, Fragments of Letters and Talks to Students, Bearing on the Concept and Technique of Picture Making, the Study of Art Generally, and on Appreciation* (Boulder, CO: Westview Press, 1984).

60. Watts, A. W. *The Way of Zen* (Penguin, 1980).

61. Endevelt-Shapira, Y., Djalovski, A., Dumas, G. & Feldman, R. Maternal chemosignals enhance infant-adult brain-to-brain synchrony. *Science Advances* 7, eabg6867 (2021).

62. Pyrkov, T. V., Avchaciov, K., Tarkhov, A. E., Menshikov, L. I., Gudkov, A. V. & Fedichev, P. O. Longitudinal analysis of blood markers reveals progressive loss of resilience and predicts human lifespan limit. *Nature Communications* 12, 1–10 (2021).

63. Pyrkov, T. V., Avchaciov, K., Tarkhov, A. E., Menshikov, L. I., Gudkov, A. V. & Fedichev, P. O. Longitudinal analysis of blood markers reveals progressive loss of resilience and predicts human lifespan limit. *Nature Communications* 12, 1–10 (2021).

64. Boehme, M., Guzzetta, K. E., Bastiaanssen, T. F., Van De Wouw, M., Moloney, G. M., Gual-Grau, A., Spichak, S., Olavarría-Ramírez, L., Fitzgerald, P., Morillas, E. & others. Microbiota from young mice counteracts selective age-associated behavioral deficits. *Nature Aging* 1, 666–76 (2021); Heijtz, R. D., Gonzalez-Santana, A. & Laman, J. D. Young microbiota rejuvenates the aging brain. *Nature Aging* 1, 625–27 (2021).

65. Grover, S., Wen, W., Viswanathan, V., Gill, C. T. & Reinhart, R. M. Long-lasting, dissociable improvements in working memory and long-term memory in older adults with repetitive neuromodulation. *Nature Neuroscience* 1–10 (2022).

66. Colman, R. J., Anderson, R. M., Johnson, S. C., Kastman, E. K., Kosmatka, K. J., Beasley, T. M., Allison, D. B., Cruzen, C., Simmons, H. A., Kemnitz, J. W. & others. Caloric restriction delays disease onset and mortality in rhesus monkeys. *Science* 325, 201–4 (2009).

67. Dias, G. P., Murphy, T., Stangl, D., Ahmet, S., Morisse, B., Nix, A., Aimone, L. J., Aimone, J. B., Kuro-O, M., Gage, F. H. & others. Intermittent fasting enhances long-term memory consolidation, adult hippocampal neurogenesis, and expression of longevity gene klotho. *Molecular Psychiatry* 26, 6365–79 (2021).

68. Hepler, C., Weidemann, B. J., Waldeck, N. J., Marcheva, B., Cedernaes, J., Thorne, A. K., Kobayashi, Y., Nozawa, R., Newman, M. V., Gao, P., Shao, M., Ramsey, K. M., Gupta, R. K. & Bass, J. Time-restricted feeding mitigates obesity through adipocyte thermogenesis. *Science* 378, 276–84 (2022).

69. Jones, C. I. *The End of Economic Growth? Unintended Consequences of a Declining Population* (Cambridge, MA: National Bureau of Economic Research, 2020); Vollset, S. E., Goren, E., Yuan, C.-W., Cao, J., Smith, A. E., Hsiao, T., Bisignano, C., Azhar, G. S., Castro, E., Chalek, J. & others. Fertility, mortality, migration, and population scenarios for 195 countries and territories from 2017 to 2100: A forecasting analysis for the global burden of disease study. *The Lancet* 396, 1285–1306 (2020).

70. Ortega, R. P. Half of Americans anticipate a US civil war soon, survey finds. *Science* 377, 357 (2022).

71. Clemm von Hohenberg, B. & Hager, A. Wolf attacks predict far-right voting. *Proceedings of the National Academy of Sciences* 119, e2202224119 (2022).

Chapter 6. Self

1. Kobayashi, M., Kikuchi, D. & Okamura, H. Imaging of ultraweak spontaneous photon emission from human body displaying diurnal rhythm. *PLoS One* 4, e6256 (2009).

2. Govinda, A., Woodroffe, J., Evans-Wentz, W. & Jung, C. *The Tibetan Book of the Dead: Or, the After-Death Experiences on the Bardo Plane, According to Lama Kazi Dawa-Samdup's English Rendering* (New York: Oxford University Press, 1960).

3. Schumacher, E. F. *A Guide for the Perplexed* (New York: Random House, 1995).

4. Govinda, A., Woodroffe, J., Evans-Wentz, W. & Jung, C. *The Tibetan Book of the Dead: Or, the After-Death Experiences on the Bardo Plane, According to Lama Kazi Dawa-Samdup's English Rendering* (New York: Oxford University Press, 1960).

5. Kocsis, B., Martínez-Bellver, S., Fiáth, R., Domonkos, A., Sviatkó, K., Schlingloff, D., Barthó, P., Freund, T. F., Ulbert, I., Káli, S. & others. Huygens synchronization of medial septal pacemaker neurons generates hippocampal theta oscillation. *Cell Reports* 40, 111149 (2022).

6. Klingler, E., Tomasello, U., Prados, J., Kebschull, J. M., Contestabile, A., Galiñanes, G. L., Fièvre, S., Santinha, A., Platt, R., Huber, D. & others. Temporal controls over inter-areal cortical projection neuron fate diversity. *Nature* 599, 453–57 (2021).

7. Hopfield, J. J. & Herz, A. V. Rapid local synchronization of action potentials: Toward computation with coupled integrate-and-fire neurons. *Proceedings of the National Academy of Sciences* 92, 6655–62 (1995).

8. Qu, Z., Hu, G., Garfinkel, A. & Weiss, J. N. Nonlinear and stochastic dynamics in the heart. *Physics Reports* 543, 61–162 (2014).

9. Michaels, D. C., Matyas, E. P. & Jalife, J. Mechanisms of sinoatrial pacemaker synchronization: A new hypothesis. *Circulation Research* 61, 704–14 (1987).

10. Qu, Z., Hu, G., Garfinkel, A. & Weiss, J. N. Nonlinear and stochastic dynamics in the heart. *Physics Reports* 543, 61–162 (2014).

11. Wiener, N. *Cybernetics: Or Control and Communication in the Animal and the Machine* (Cambridge, MA: MIT Press, 1948), 212.

12. Lovelock, J. *Gaia: A New Look at Life on Earth* (Oxford, UK: Oxford University Press, 1995), 148.

13. Watts, A. W. *The Way of Zen* (New York: Penguin, 1980).

14. Bowles, S., Hickman, J., Peng, X., Williamson, W. R., Huang, R., Washington, K., Donegan, D. & Welle, C. G. Vagus nerve stimulation drives selective circuit modulation through cholinergic reinforcement. *Neuron* 110, 2867–85 (2022).

15. Edwards, C. A., Kouzani, A., Lee, K. H. & Ross, E. K. Neurostimulation devices for the treatment of neurologic disorders. *Mayo Clinic Proceedings* 92, 1427–44 (2017).

16. Narvaez, D., Wang, L., Cheng, A., Gleason, T. R., Woodbury, R., Kurth, A. & Lefever, J. B. The importance of early life touch for psychosocial and moral development. *Psicologia: Reflexão e Crítica* 32 (2019).

17. Eyre, M., Fitzgibbon, S. P., Ciarrusta, J., Cordero-Grande, L., Price, A. N., Poppe, T., Schuh, A., Hughes, E., O'Keeffe, C., Brandon, J. & others. The developing human connectome project: Typical and disrupted perinatal functional connectivity. *Brain* 144, 2199–213 (2021).

18. Kelsey, C. M., Farris, K. & Grossmann, T. Variability in infants' functional brain network connectivity is associated with differences in affect and behavior. *Frontiers in Psychiatry* 12, 896 (2021).

19. Huszár, R., Zhang, Y., Blockus, H. & Buzsáki, G. Preconfigured dynamics in the hippocampus are guided by embryonic birthdate and rate of neurogenesis. *bioRxiv* (2022).

20. Chen, P.-C., Niknazar, H., Alaynick, W. A., Whitehurst, L. N. & Mednick, S. C. Competitive dynamics underlie cognitive improvements during sleep. *Proceedings of the National Academy of Sciences* 118, e2109339118 (2021).

21. Owen, L. L., Chang, T. H. & Manning, J. R. High-level cognition during story listening is reflected in high-order dynamic correlations in neural activity patterns. *Nature Communications* 12, 1–14 (2021).

22. Donoghue, T., Haller, M., Peterson, E. J., Varma, P., Sebastian, P., Gao, R., Noto, T., Lara, A. H., Wallis, J. D., Knight, R. T. & others. Parameterizing neural power spectra into periodic and aperiodic components. *Nature Neuroscience* 23, 1655–65 (2020).

23. Tagliazucchi, E., Balenzuela, P., Fraiman, D. & Chialvo, D. R. Criticality in large-scale brain fMRI dynamics unveiled by a novel point process analysis. *Frontiers in Physiology* 3, 15 (2012).

24. Seth, A. *Being You: A New Science of Consciousness* (New York: Penguin, 2021).

25. Ge, X., Zhang, K., Gribizis, A., Hamodi, A. S., Sabino, A. M. & Crair, M. C. Retinal waves prime visual motion detection by simulating future optic flow. *Science* 373, eabd0830 (2021).

26. Seth, A. *Being You: A New Science of Consciousness* (New York: Penguin, 2021).

27. Henri, R. *The Art Spirit: Notes, Articles, Fragments of Letters and Talks to Students, Bearing on the Concept and Technique of Picture Making, the Study of Art Generally, and on Appreciation* (Westview Press, 1984).

28. Gasquet, J., and Cézanne P. *Joachim Gasquet's Cézanne, a Memoir with Conversations* (London: Thames and Hudson, 1991).

29. Capra, F. *The Hidden Connections: A Science for Sustainable Living* (Anchor, 2004).

30. Wilczek, F. A. *Beautiful Question* (Penguin Random House, 2016), 430.

31. Seth, A. *Being You: A New Science of Consciousness* (Penguin, 2021).

32. Hansen, N. C., Kragness, H. E., Vuust, P., Trainor, L. & Pearce, M. T. Predictive uncertainty underlies auditory boundary perception. *Psychological Science* 32, 1416–25 (2021).

33. Wohltjen, S. & Wheatley, T. Eye contact marks the rise and fall of shared attention in conversation. *Proceedings of the National Academy of Sciences* 118, e2106645118 (2021).

34. Song, H., Finn, E. S. & Rosenberg, M. D. Neural signatures of attentional engagement during narratives and its consequences for event memory. *Proceedings of the National Academy of Sciences* 118, e2021905118 (2021).

35. Hamilton, L. S., Oganian, Y., Hall, J. & Chang, E. F. Parallel and distributed encoding of speech across human auditory cortex. *Cell* 184, 4626–39 (2021).

36. Gold, B. P., Pearce, M. T., Mas-Herrero, E., Dagher, A. & Zatorre, R. J. Predictability and uncertainty in the pleasure of music: A reward for learning? *Journal of Neuroscience* 39, 9397–409 (2019); Isik, A. I. & Vessel, E. A. From visual perception to aesthetic appeal: Brain responses to aesthetically appealing natural landscape movies. *Frontiers in Human Neuroscience* 414 (2021).

37. Twenge, J. M. & Manis, M. First-name desirability and adjustment: Self-satisfaction, others' ratings, and family background. *Journal of Applied Social Psychology* 28, 41–51 (1998).

38. Bao, H.-W.-S., Cai, H., DeWall, C. N., Gu, R., Chen, J., Luo, Y. L. & others. Name uniqueness predicts career choice and career achievement. *PsyArXiv* (2020).

39. Kang, Y., Zhu, D. H. & Zhang, Y. A. Being extraordinary: How CEOs' uncommon names explain strategic distinctiveness. *Strategic Management Journal* 42, 462–88 (2021).

40. Krakauer, D., Bertschinger, N., Olbrich, E., Flack, J. C. & Ay, N. The information theory of individuality. *Theory in Biosciences* 139, 209–23 (2020).

41. Johnstone, B., Cohen, D. & Dennison, A. The integration of sensations and mental experiences into a unified experience: A neuropsychological model for the "sense of self." *Neuropsychologia* 159, 107939 (2021).

42. Xu, X., Yuan, H. & Lei, X. Activation and connectivity within the default mode network contribute independently to future-oriented thought. *Scientific Reports* 6, 1–10 (2016).

43. Sambataro, F., Wolf, N. D., Giusti, P., Vasic, N. & Wolf, R. C. Default mode network in depression: A pathway to impaired affective cognition? *Clinical Neuropsychiatry* 10 (2013).

44. Raichle, M. E., MacLeod, A. M., Snyder, A. Z., Powers, W. J., Gusnard, D. A. & Shulman, G. L. A default mode of brain function. *Proceedings of the National Academy of Sciences* 98, 676–82 (2001).

45. Andrews-Hanna, J. R., Smallwood, J. & Spreng, R. N. The default network and self-generated thought: Component processes, dynamic control, and clinical relevance. *Annals of the New York Academy of Sciences* 1316, 29–52 (2014).

46. Vessel, E. A., Starr, G. G. & Rubin, N. The brain on art: Intense aesthetic experience activates the default mode network. *Frontiers in Human Neuroscience* 6, 66 (2012); Vessel, E. A., Starr, G. G. & Rubin, N. Art reaches within: Aesthetic experience, the self and the default mode network. *Frontiers in Neuroscience* 7, 258 (2013).

47. Perkins, A. M., Arnone, D., Smallwood, J. & Mobbs, D. Thinking too much: Self-generated thought as the engine of neuroticism. *Trends in Cognitive Sciences* 19, 492–98 (2015).

48. Hari, J. *Lost Connections: Why You're Depressed and How to Find Hope* (New York: Bloomsbury, 2019).

49. Nikolova, V. L., Hall, M. R., Hall, L. J., Cleare, A. J., Stone, J. M. & Young, A. H. Perturbations in gut microbiota composition in psychiatric disorders: A review and meta-analysis. *JAMA Psychiatry* 78, 1343–54 (2021).

50. Brüchle, W., Schwarzer, C., Berns, C., Scho, S., Schneefeld, J., Koester, D., Schack, T., Schneider, U. & Rosenkranz, K. Physical activity reduces clinical symptoms and restores neuroplasticity in major depression. *Frontiers in Psychiatry* 935 (2021).

51. Moncrieff, J., Cooper, R. E., Stockmann, T., Amendola, S., Hengartner, M. P. & Horowitz, M. A. The serotonin theory of depression: A systematic umbrella review of the evidence. *Molecular Psychiatry* 1–14 (2022).

52. Lafrance, A., Strahan, E., Bird, B. M., St. Pierre, M. & Walsh, Z. Classic psychedelic use and mechanisms of mental health: Exploring the mediating roles of spirituality and emotion processing on symptoms of anxiety, depressed mood, and disordered eating in a community sample. *Journal of Humanistic Psychology* 00221678211048049 (2021).

53. Klein, A. S., Dolensek, N., Weiand, C. & Gogolla, N. Fear balance is maintained by bodily feedback to the insular cortex in mice. *Science* 374, 1010–15 (2021).

54. Schmitz, L. L. & Duque, V. In utero exposure to the Great Depression is reflected in late-life epigenetic aging signatures. *Proceedings of the National Academy of Sciences* 119, e2208530119 (2022).

55. Riccardo, B., Kanat, C., Michele, P., Li, X., Simon, S., Esi, D., Gaelle, A., Andrea, C., Wiskerke, J., Szczot, I. & others. An epigenetic mechanism for over-consolidation of fear memories. *Molecular Psychiatry* 1–12 (2022).

56. Zanette, L. Y. & Clinchy, M. Ecology and neurobiology of fear in free-living wildlife. *Annual Review of Ecology, Evolution, and Systematics* 51, 297–318 (2020).

57. Seth, A. *Being You: A New Science of Consciousness* (New York: Penguin, 2021).

Chapter 7. Beyond

1. West, G. B., Brown, J. H. & Enquist, B. J. A general model for the structure and allometry of plant vascular systems. *Nature* 400, 664–67 (1999).

2. Simard, S. W., Beiler, K. J., Bingham, M. A., Deslippe, J. R., Philip, L. J. & Teste, F. P. Mycorrhizal networks: Mechanisms, ecology and modeling. *Fungal Biology Reviews* 26, 39–60 (2012).

3. Simard, S. *Finding the Mother Tree: Uncovering the Wisdom and Intelligence of the Forest* (London: Penguin, 2021).

4. Valencia, A. L. & Froese, T. What binds us? Inter-brain neural synchronization and its implications for theories of human consciousness. *Neuroscience of Consciousness* 2020, niaa010 (2020).

5. Seth, A. *Being You: A New Science of Consciousness* (New York: Penguin, 2021).

6. Brach, T. *Radical Acceptance: Embracing Your Life with the Heart of a Buddha* (New York: Bantam, 2004).

7. Ginsburg, S. & Jablonka, E. *The Evolution of the Sensitive Soul: Learning and the Origins of Consciousness* (Cambridge: MIT Press, 2019).

8. Tononi, G. An information integration theory of consciousness. *BMC Neuroscience* 5, 1–22 (2004).

9. Jost, J. Information theory and consciousness. *Frontiers in Applied Mathematics and Statistics* 50 (2021).

10. Budson, A. E., Richman, K. A. & Kensinger, E. A. Consciousness as a memory system. *Cognitive and Behavioral Neurology* 35 (2022).

11. Budson, A. E., Richman, K. A. & Kensinger, E. A. Consciousness as a memory system. *Cognitive and Behavioral Neurology* 35 (2022).

12. Tononi, G. An information integration theory of consciousness. *BMC Neuroscience* 5, 1–22 (2004).

13. Seth, A. *Being You: A New Science of Consciousness* (New York: Penguin, 2021).

14. Mas-Herrero, E., Dagher, A., Farrés-Franch, M. & Zatorre, R. J. Unraveling the temporal dynamics of reward signals in music-induced pleasure with TMS. *Journal of Neuroscience* 41, 3889–99 (2021).

15. Leipold, S., Klein, C. & Jäncke, L. Musical expertise shapes functional and structural brain networks independent of absolute pitch ability. *Journal of Neuroscience* 41, 2496–511 (2021).

16. Safron, A. What is orgasm? A model of sexual trance and climax via rhythmic entrainment. *Socioaffective Neuroscience & Psychology* 6, 31763 (2016).

17. Govinda, A., Woodroffe, J., Evans-Wentz, W. & Jung, C. *The Tibetan Book of the Dead: Or, the After-Death Experiences on the Bardo Plane, According to Lama Kazi Dawa-Samdup's English Rendering* (Oxford, UK: Oxford University Press, 1960).

18. Rudgley, R. *The Alchemy of Culture: Intoxicants in Society* (London: British Museum Press, 1993).

19. Ballentine, G., Friedman, S. F. & Bzdok, D. Trips and neurotransmitters: Discovering principled patterns across 6850 hallucinogenic experiences. *Science Advances* 8, eabl6989 (2022).

20. Seth, A. *Being You: A New Science of Consciousness* (New York: Penguin, 2021).

21. Maté, G. *The Myth of Normal: Trauma, Illness, and Healing in a Toxic Culture* (New York: Avery, 2022).

22. Ballentine, G., Friedman, S. F. & Bzdok, D. Trips and neurotransmitters: Discovering principled patterns across 6850 hallucinogenic experiences. *Science Advances* 8, eabl6989 (2022).

23. Carhart-Harris, R. L., Muthukumaraswamy, S., Roseman, L., Kaelen, M., Droog, W., Murphy, K., Tagliazucchi, E., Schenberg, E. E., Nest, T., Orban, C. & others. Neural correlates of the LSD experience revealed by multimodal neuroimaging. *Proceedings of the National Academy of Sciences* 113, 4853–58 (2016).

24. Shao, L.-X., Liao, C., Gregg, I., Davoudian, P. A., Savalia, N. K., Delagarza, K. & Kwan, A. C. Psilocybin induces rapid and persistent growth of dendritic spines in frontal cortex in vivo. *Neuron* 109, 2535–44 (2021).

25. Daws, R. E., Timmermann, C., Giribaldi, B., Sexton, J. D., Wall, M. B., Erritzoe, D., Roseman, L., Nutt, D. & Carhart-Harris, R. Increased global integration in the brain after psilocybin therapy for depression. *Nature Medicine* 28, 844–51 (2022).

26. Ballentine, G., Friedman, S. F. & Bzdok, D. Trips and neurotransmitters: Discovering principled patterns across 6850 hallucinogenic experiences. *Science Advances* 8, eabl6989 (2022).

27. Weiss, B., Miller, J. D., Carter, N. T. & Keith Campbell, W. Examining changes in personality following shamanic ceremonial use of ayahuasca. *Scientific Reports* 11, 1–15 (2021).

28. Nayak, S. M. & Griffiths, R. R. A single belief-changing psychedelic experience is associated with increased attribution of consciousness to living and non-living entities. *Frontiers in Psychology* 1035 (2022).

29. Cohen, D. *The Evolution of Religion, Religiosity and Theology* (London: Routledge, 2020), 54–69.

30. Ihm, E. D., Paloutzian, R. F., Elk, M. van & Schooler, J. W. *The Evolution of Religion, Religiosity and Theology* (London: Routledge, 2020), 138–53.

31. Capra, F. *The Hidden Connections: A Science for Sustainable Living* (New York: Anchor, 2004).

32. Hitchcock, R. K. *The Evolution of Religion, Religiosity and Theology* (London: Routledge, 2020), 239–55.

33. Narby, J. *The Cosmic Serpent* (London: Phoenix, 1998).

34. Niggli, H. J. & Applegate, L. A. *Integrative Biophysics* (Dordrecht: Springer, 2003), 361–85.

35. Popp, F.-A. *Integrative Biophysics* (Dordrecht: Springer, 2003), 387–438.

36. Broom, D. M. *The Evolution of Religion, Religiosity and Theology* (London: Routledge, 2020), 70–84.

37. Tesla, N. *My Inventions and Other Writings* (Digireads, 2017).

38. Banik, I. & Zhao, H. From galactic bars to the Hubble tension: Weighing up the astrophysical evidence for Milgromian gravity. *Symmetry* 14, 1331 (2022).

39. Hunter, R. L. *Plato's Symposium* (Oxford, UK: Oxford University Press on Demand, 2004).

40. Ferguson, M. A., Schaper, F. L., Cohen, A., Siddiqi, S., Merrill, S. M., Nielsen, J. A., Grafman, J., Urgesi, C., Fabbro, F. & Fox, M. D. A neural circuit for spirituality and religiosity derived from patients with brain lesions. *Biological Psychiatry* 91, 380–88 (2022).

41. Miller, L. *The Awakened Brain* (New York: Random House, 2022).

42. Dyer, W. W. *You'll See It When You Believe It: The Way to Your Personal Transformation* (New York: William Morrow, 1989).

43. Miller, L. *The Awakened Brain* (New York: Random House, 2022).

44. Wahbeh, H., Sagher, A., Back, W., Pundhir, P. & Travis, F. A systematic review of transcendent states across meditation and contemplative traditions. *Explore* 14, 19–35 (2018).

45. Jiang, H., He, B., Guo, X., Wang, X., Guo, M., Wang, Z., Xue, T., Li, H., Xu, T., Ye, S. & others. Brain-heart interactions underlying traditional Tibetan Buddhist meditation. *Cerebral Cortex* 30, 439–50 (2020).

46. Gamma, A. & Metzinger, T. The Minimal Phenomenal Experience questionnaire (MPE-92M): Towards a phenomenological profile of "pure awareness" experiences in meditators. *PLOS One* 16, e0253694 (2021).

47. Wahbeh, H., Sagher, A., Back, W., Pundhir, P. & Travis, F. A systematic review of transcendent states across meditation and contemplative traditions. *Explore* 14, 19–35 (2018).

48. Wahbeh, H., Sagher, A., Back, W., Pundhir, P. & Travis, F. A systematic review of transcendent states across meditation and contemplative traditions. *Explore* 14, 19–35 (2018).

49. Bremer, B., Wu, Q., Mora Álvarez, M. G., Hölzel, B. K., Wilhelm, M., Hell, E., Tavacioglu, E. E., Torske, A. & Koch, K. Mindfulness meditation increases default mode, salience, and central executive network connectivity. *Scientific Reports* 12, 1–15 (2022).

50. Fox, K. C., Nijeboer, S., Dixon, M. L., Floman, J. L., Ellamil, M., Rumak, S. P., Sedlmeier, P. & Christoff, K. Is meditation associated with altered brain structure? A systematic review and meta-analysis of morphometric neuroimaging in meditation practitioners. *Neuroscience & Biobehavioral Reviews* 43, 48–73 (2014).

51. Riegner, G., Posey, G., Oliva, V., Jung, Y., Mobley, W. & Zeidan, F. Disentangling self from pain: Mindfulness meditation-induced pain relief is driven by thalamic-default mode network decoupling. *Pain* (2022).

52. Chandran, V., Bermúdez, M.-L., Koka, M., Chandran, B., Pawale, D., Vishnubhotla, R., Alankar, S., Maturi, R., Subramaniam, B. & Sadhasivam, S. Large-scale genomic study reveals robust activation of the immune system following advanced inner engineering meditation retreat. *Proceedings of the National Academy of Sciences* 118, e2110455118 (2021).

53. Jedamzik, K. & Pogosian, L. Relieving the Hubble tension with primordial magnetic fields. *Physical Review Letters* 125, 181302 (2020).

54. Harrison, E. Origin of magnetic fields in the early universe. *Physical Review Letters* 30, 188 (1973).

55. Tully, R. B., Courtois, H., Hoffman, Y. & Pomarède, D. The Laniakea supercluster of galaxies. *Nature* 513, 71–73 (2014).

56. Wang, P., Libeskind, N. I., Tempel, E., Kang, X. & Guo, Q. Possible observational evidence for cosmic filament spin. *Nature Astronomy* 5, 839–45 (2021).

57. Ruderman, M. A. Possible consequences of nearby supernova explosions for atmospheric ozone and terrestrial life. *Science* 184, 1079–81 (1974).

58. Ellis, J. & Schramm, D. N. Could a nearby supernova explosion have caused a mass extinction? *Proceedings of the National Academy of Sciences* 92, 235–38 (1995); Melott, A. L., Marinho, F. & Paulucci, L. Hypothesis: Muon radiation dose and marine megafaunal extinction at the end-Pliocene supernova. *Astrobiology* 19, 825–30 (2019).

59. Hameroff, S. & Penrose, R. Consciousness in the universe: A review of the "Orch OR" theory. *Physics of Life Reviews* 11, 39–78 (2014).

60. Alexander, S., Cunningham, W. J., Lanier, J., Smolin, L., Stanojevic, S., Toomey, M. W. & Wecker, D. The autodidactic universe. *arXiv preprint* arXiv:2104.03902 (2021).

61. Vopson, M. M. The mass-energy-information equivalence principle. *AIP Advances* 9, 095206 (2019).

62. Sharma, A., Czégel, D., Lachmann, M., Kempes, C. P., Walker, S. I. & Cronin, L. Assembly theory explains and quantifies the emergence of selection and evolution. *arXiv preprint* arXiv:2206.02279 (2022).

63. Vince, G. *Transcendence: How Humans Evolved through Fire, Language, Beauty, and Time* (London: Penguin, 2019).

Chapter 8. Synthesis

1. Kondepudi, D., Kay, B. & Dixon, J. End-directed evolution and the emergence of energy-seeking behavior in a complex system. *Physical Review E* 91, 050902 (2015).

2. England, J. L. Dissipative adaptation in driven self-assembly. *Nature Nanotechnology* 10, 919–23 (2015).

3. Sheldrake, M. *Entangled Life: How Fungi Make Our Worlds, Change Our Minds & Shape Our Futures* (New York: Random House, 2020).

4. Kienle, K., Glaser, K. M., Eickhoff, S., Mihlan, M., Knöpper, K., Reátegui, E., Epple, M. W., Gunzer, M., Baumeister, R., Tarrant, T. K. & others. Neutrophils self-limit swarming to contain bacterial growth in vivo. *Science* 372, eabe7729 (2021).

5. Bak, P. & Chen, K. Self-organized criticality. *Scientific American* 264, 46–53 (1991).

6. Strogatz, S. H. *Sync: How Order Emerges from Chaos in the Universe, Nature, and Daily Life* (New York: Hachette, 2012).

7. Capra, F. *The Hidden Connections: A Science for Sustainable Living* (New York: Anchor, 2004).

8. England, J. L. Dissipative adaptation in driven self-assembly. *Nature Nanotechnology* 10, 919–23 (2015).

9. Speed, H. *The Practice and Science of Drawing* (Seeley, Service and Co., 1913).

10. Speer, M. E., Ibrahim, S., Schiller, D. & Delgado, M. R. Finding positive meaning in memories of negative events adaptively updates memory. *Nature Communications* 12, 1–11 (2021).

11. Clay, G., Mlynski, C., Korb, F. M., Goschke, T. & Job, V. Rewarding cognitive effort increases the intrinsic value of mental labor. *Proceedings of the National Academy of Sciences* 119, e2111785119 (2022).

12. Papers of John F. Kennedy. Presidential Papers. Address at Rice University, Houston, Texas, 12 September 1962. www.jfklibrary.org/asset-viewer/archives/JFKPOF/040/JFKPOF-040-001.

13. Maturana, H. R. & Varela, F. J. *Autopoiesis and Cognition: The Realization of the Living* (Dordrecht: Springer, 1991).

14. Kempes, C. P. & Krakauer, D. C. The multiple paths to multiple life. *Journal of Molecular Evolution* 89, 415–26 (2021).

15. Johnston, I. G., Dingle, K., Greenbury, S. F., Camargo, C. Q., Doye, J. P., Ahnert, S. E. & Louis, A. A. Symmetry and simplicity spontaneously emerge from the algorithmic nature of evolution. *Proceedings of the National Academy of Sciences* 119, e2113883119 (2022).

16. Sharma, A., Czégel, D., Lachmann, M., Kempes, C. P., Walker, S. I. & Cronin, L. Assembly theory explains and quantifies the emergence of selection and evolution. *arXiv preprint*, arXiv:2206.02279 (2022).

17. Taylor, H., Fernandes, B. & Wraight, S. The evolution of complementary cognition: Humans cooperatively adapt and evolve through a system of collective cognitive search. *Cambridge Archaeological Journal* 32, 61–77 (2022).

18. Frank, A., Grinspoon, D. & Walker, S. Intelligence as a planetary scale process. *International Journal of Astrobiology* 21, 47–61 (2022).

19. Golan, A. & Harte, J. Information theory: A foundation for complexity science. *Proceedings of the National Academy of Sciences* 119, e2119089119 (2022).

20. Notarmuzi, D., Castellano, C., Flammini, A., Mazzilli, D. & Radicchi, F. Universality, criticality and complexity of information propagation in social media. *Nature Communications* 13, 1–8 (2022).

21. Barabasi, A.-L. The origin of bursts and heavy tails in human dynamics. *Nature* 435, 207–11 (2005).

22. Koster, R., Balaguer, J., Tacchetti, A., Weinstein, A., Zhu, T., Hauser, O., Williams, D., Campbell-Gillingham, L., Thacker, P., Botvinick, M. & others. Human-centered mechanism design with democratic AI. *Nature Human Behaviour* https://doi.org/10.1038/s41562-022-01383-x (2022).

23. Speed, H. *The Practice and Science of Drawing* (Seeley, Service and Co., 1913).

24. Schumacher, E. F. *A Guide for the Perplexed* (New York: Random House, 1995).

25. Nietzsche, F. *The Birth of Tragedy out of the Spirit of Music*, trans. S. Whiteside, ed. Michael Tanner (London: Penguin, 1993).

26. Seth, A. *Being You: A New Science of Consciousness* (New York: Penguin, 2021).

27. Gabriel, R. Affect, belief, and the arts. *Frontiers in Psychology* 12 (2021).

28. Gabriel, R. Affect, belief, and the arts. *Frontiers in Psychology* 12 (2021).

29. Wilczek, F. *A Beautiful Question* (New York: Penguin, 2016), 430.

30. Henri, R. *The Art Spirit: Notes, Articles, Fragments of Letters and Talks to Students, Bearing on the Concept and Technique of Picture Making, the Study of Art Generally, and on Appreciation* (Boulder, CO: Westview Press, 1984).

31. Schumacher, E. F. *A Guide for the Perplexed* (New York: Random House, 1995).

32. Schumacher, E. F. *A Guide for the Perplexed* (New York: Random House, 1995).

33. Govinda, A., Woodroffe, J., Evans-Wentz, W. & Jung, C. *The Tibetan Book of the Dead: Or, the After-Death Experiences on the Bardo Plane, According to Lama Kazi Dawa-Samdup's English Rendering* (Oxford, UK: Oxford University Press, 1960).

INDEX